Series in Astronomy and Astrophysics

Very High Energy Gamma-Ray Astronomy

Trevor Weekes

Whipple Observatory, Harvard–Smithsonian Center for Astrophysics, USA

Institute of Physics Publishing
Bristol and Philadelphia

© IOP Publishing Ltd 2003

All rights reserved. No part of this publication may be reproduced, stored in a retrieval system or transmitted in any form or by any means, electronic, mechanical, photocopying, recording or otherwise, without the prior permission of the publisher. Multiple copying is permitted in accordance with the terms of licences issued by the Copyright Licensing Agency under the terms of its agreement with Universities UK (UUK).

Trevor Weekes has asserted his moral right under the Copyright, Designs and Patents Act 1998 to be identified as the author of this work.
British Library Cataloguing-in-Publication Data
A catalogue record for this book is available from the British Library.

ISBN 0 7503 0658 0

Library of Congress Cataloging-in-Publication Data are available

Series Editors: **M Birkinshaw**, University of Bristol, UK
M Elvis, Harvard–Smithsonian Center for Astrophysics, USA
J Silk, University of Oxford, UK

Commissioning Editor: John Navas
Production Editor: Simon Laurenson
Production Control: Sarah Plenty
Cover Design: Victoria Le Billon
Marketing: Nicola Newey and Verity Cooke

Published by Institute of Physics Publishing, wholly owned by The Institute of Physics, London

Institute of Physics Publishing, Dirac House, Temple Back, Bristol BS1 6BE, UK

US Office: Institute of Physics Publishing, The Public Ledger Building, Suite 929, 150 South Independence Mall West, Philadelphia, PA 19106, USA

Typeset in $\LaTeX\,2_\varepsilon$ by Text 2 Text, Torquay, Devon
Printed in the UK by MPG Books Ltd, Bodmin, Cornwall

Very High Energy Gamma-Ray Astronomy

Series in Astronomy and Astrophysics

Series Editors: **M Birkinshaw**, University of Bristol, UK
M Elvis, Harvard–Smithsonian Center for Astrophysics, USA
J Silk, University of Oxford, UK

The Series in Astronomy and Astrophysics includes books on all aspects of theoretical and experimental astronomy and astrophysics. Books in the series range in level from textbooks and handbooks to more advanced expositions of current research.

Other books in the series

The Physics of Interstellar Dust
E Krügel

Dark Sky, Dark Matter
J M Overduin and P S Wesson

Dust in the Galactic Environment, 2nd Edition
D C B Whittet

An Introduction to the Science of Cosmology
D J Raine and E G Thomas

The Origin and Evolution of the Solar System
M M Woolfson

The Physics of the Interstellar Medium
J E Dyson and D A Williams

Dust and Chemistry in Astronomy
T J Millar and D A Williams (eds)

Observational Astrophysics
R E White (ed)

Stellar Astrophysics
R J Tayler (ed)

Forthcoming titles

Numerical Methods in Astrophysics
P Bodenheimer, G Laughlin, M Rozyczka and H W Yorke

To Ann who gave me moral support through four decades of gamma-ray astronomy

Contents

	Foreword		**xiii**
1	**Foundations of gamma-ray astronomy**		**1**
	1.1	Astronomical exploration	1
	1.2	The relativistic universe	2
	1.3	Definitions	4
	1.4	The heroic era of gamma-ray astronomy	5
		1.4.1 The early promise	5
		1.4.2 Peculiarities of gamma-ray telescopes	6
		1.4.3 VHE gamma-ray telescopes on the ground	7
		Historical note: seminal paper	10
		1.4.4 HE gamma-ray telescopes in space	11
2	**Very high energy gamma-ray detectors**		**13**
	2.1	The atmospheric windows	13
	2.2	Electromagnetic cascade in atmosphere	14
	2.3	The visible electromagnetic cascade	14
	2.4	Atmospheric Cherenkov technique	18
		2.4.1 General properties	18
		2.4.2 Features of the technique	21
	2.5	The background of cosmic radiation	25
		2.5.1 Charged cosmic rays	25
		2.5.2 Flux sensitivity	27
	2.6	Atmospheric Cherenkov imaging detectors	28
		2.6.1 Principle	28
		2.6.2 Angular resolution	30
		2.6.3 Energy resolution	30
		2.6.4 Existing imaging telescopes	30
		2.6.5 Arrays	34
	2.7	Other ground-based detectors	38
		2.7.1 Particle air shower arrays	38
		2.7.2 Solar power stations as ACTs	38
		Historical note: Cherenkov images	40

viii   Contents

3 High energy gamma-ray telescopes in space — **42**
- 3.1 Introduction — 42
- 3.2 Pair production telescopes: high energy — 42
- 3.3 Compton telescopes — 46
- 3.4 Future space telescopes — 48
  - 3.4.1 INTEGRAL — 48
  - 3.4.2 Swift — 49
  - 3.4.3 Light imaging detector for gamma-ray astronomy (AGILE) — 49
  - 3.4.4 Alpha Magnetic Spectrometer (AMS) — 50
  - 3.4.5 The Gamma-ray Large-Area Space Telescope (GLAST) — 50
- Historical note: CGRO rescue — 54

4 Galactic plane — **55**
- 4.1 Study of the galactic plane — 55
- 4.2 Gamma-ray observations — 58
  - 4.2.1 HE observations — 58
  - 4.2.2 VHE observations — 58
- 4.3 Interpretation — 60
- 4.4 Energy spectrum — 62
- Historical note — 65

5 Supernovae and supernova remnants — **67**
- 5.1 Supernova explosions — 67
- 5.2 Energy considerations — 68
- 5.3 Acceleration — 70
- 5.4 Detection at outburst — 71
- 5.5 Supernova remnant classification — 72
- 5.6 SNRs as cosmic ray sources — 74
- Historical note: SN1987a — 75

6 Gamma-ray observations of the Crab Nebula — **77**
- 6.1 Significance — 77
- 6.2 Optical and x-ray observations — 78
- 6.3 Gamma-ray history — 79
  - 6.3.1 HE observations — 79
  - 6.3.2 VHE observations — 82
- 6.4 Gamma source — 83
  - 6.4.1 The Crab resolved — 83
  - 6.4.2 The standard candle — 85
  - 6.4.3 Interpretation — 88
- Historical box: Crab pictograph — 90

7	**Gamma-ray observations of supernova remnants**		**92**
	7.1	Introduction	92
	7.2	Plerions	92
		7.2.1 SNR/PSR1706-44	92
		7.2.2 Vela	93
	7.3	Shell-type SNRs	93
		7.3.1 SN1006	93
		7.3.2 RXJ1713.7-3946	95
		7.3.3 Cassiopeia A	95
		7.3.4 Other possible detections	97
		Historical note: supernova of 1006	99
8	**Gamma-ray pulsars and binaries**		**102**
	8.1	General properties of pulsars	102
	8.2	Gamma-ray observations	104
		8.2.1 General characteristics	104
		8.2.2 Spectral energy distribution	105
		8.2.3 Light curves	106
	8.3	Models	109
		8.3.1 Polar cap models	109
		8.3.2 Outer gap models	109
	8.4	Outlook	111
	8.5	Binaries	111
		Historical note: Cygnus X-3	113
9	**Unidentified sources**		**116**
	9.1	HE observations	116
	9.2	Population studies	117
	9.3	Individual identifications	120
		9.3.1 CG135+01	120
		9.3.2 3EG J0634+0521: binary pulsar?	121
		9.3.3 3EG J1835+5918: Geminga-like pulsar?	121
		9.3.4 Galactic center	121
	9.4	Microquasars	122
	9.5	VHE observations	123
		Historical note: Geminga	124
10	**Extragalactic sources**		**126**
	10.1	Introduction	126
	10.2	Galaxies: classification	126
	10.3	Normal galaxies	127
	10.4	Starburst galaxies	128
	10.5	Active galaxies	128
		10.5.1 Radio galaxies	129
		10.5.2 Active galactic nuclei	130

x Contents

  Historical note: cosmic ray origins   133

11 Active galactic nuclei: observations   **134**
 11.1 Gamma-ray blazars   134
 11.2 Gamma-ray observations: HE   134
  11.2.1 HE source catalog   134
  11.2.2 Distance   135
  11.2.3 Classification   135
  11.2.4 Time variability   135
  11.2.5 Luminosity   137
  11.2.6 Spectrum   137
  11.2.7 Multi-wavelength observations   137
  11.2.8 Spectral energy distributions   139
  11.2.9 Future prospects   140
 11.3 Gamma-ray observations: VHE   140
  11.3.1 VHE source catalog   140
  11.3.2 Distance   143
  11.3.3 Classification   143
  11.3.4 Variability   143
  11.3.5 Luminosity   144
  11.3.6 Spectrum   144
  11.3.7 Multi-wavelength observations   148
  11.3.8 Spectral energy distributions   149
  11.3.9 Future prospects   154
  Historical note: discovery of 3C279   154

12 Active galactic nuclei: models   **156**
 12.1 Phenomenon   156
 12.2 Source of energy   157
 12.3 Beaming   158
 12.4 Models   159
  12.4.1 Lepton models   160
  12.4.2 Proton models   162
 12.5 Implications of the gamma-ray observations   163
  12.5.1 HE observations   163
  12.5.2 VHE observations   164
  12.5.3 Unified theories   166
  Historical note: superluminal motion   166

13 Gamma-ray bursts   **169**
 13.1 Introduction   169
 13.2 The discovery   169
 13.3 Properties of gamma-ray bursts   175
  13.3.1 Time profiles   175
  13.3.2 Energy spectra   175

	13.3.3	Intensity distribution	176
	13.3.4	Distribution of arrival directions	176
13.4		The location controversy	177
13.5		Counterparts	179
13.6		The high energy component	182
13.7		The afterglow	185
13.8		Models	186
	13.8.1	Central engine	186
	13.8.2	Total energies	186
	13.8.3	Beaming	187
	13.8.4	Emission mechanism	187
	13.8.5	Geometry	187
		Historical note: the great debate	188

14 Diffuse background radiation — 190

- 14.1 Measurement difficulties — 190
- 14.2 Diffuse gamma-ray background — 191
  - 14.2.1 Observations — 191
  - 14.2.2 Interpretation — 191
- 14.3 Extragalactic background light — 194
  - 14.3.1 Stellar connection — 194
  - 14.3.2 Measurement of the soft EBL — 195
  - 14.3.3 VHE observations — 197
  - Historical note: the 1 MeV bump — 198

Appendix: Radiation and absorption processes — 200

- A.1 Introduction — 200
- A.2 Compton scattering — 200
- A.3 Pair production — 205
- A.4 Electron bremsstrahlung — 206
- A.5 Pion production — 207
- A.6 Gamma-ray absorption — 208
  - A.6.1 Pair production on matter — 208
  - A.6.2 Photon–photon pair production — 208
- A.7 Synchrotron radiation — 209
- A.8 Cherenkov radiation — 211
- Historical note: distance limit — 214

Index — 217

Foreword

Astronomy is a conservative branch of science and astronomers have not always been quick to acknowledge and to welcome new avenues of research for the investigation of cosmic sources. This is particularly true when the new discipline is limited to a small number of, possibly pathological, objects. Radio astronomy was slow to be accepted because it was soon apparent that most stars were not radio sources. In contrast, x-ray astronomy, once the techniques were sufficiently developed, was immediately recognized as a true 'astronomy' since almost every star and galaxy, at some level, was seen to be an x-ray emitter. With the advent of detailed spectral and imaging techniques, it was quickly seen that x-ray astronomers, even if their detectors and observatories were strange, still spoke the language of astronomy.

This is not the case with gamma-ray astronomy. Gamma-ray sources, particularly high energy ones, are as sparse in the cosmos as they are on earth. The telescopes used to detect them are unlike those in any other waveband and there is a complete absence of gamma-ray reflecting optics: the 'telescope' is only as big as the detector! The actual detectors have more in common with particle physics laboratories than astronomical observatories and the practitioners have generally come from the high energy particle physics community. It is small wonder, therefore, that the astronomical community has been reluctant to consider gamma-ray astronomy as a legitimate or useful discipline for astronomical investigation. The fact that the early history of gamma-ray astronomy was muddied by over-enthusiastic interpretation of marginal results did not help.

As the techniques have been developed and detections put on a firm footing, it has become apparent that it is the highest energy photons that are the real tests of source models. The detection of gamma-ray bursts has opened the eyes of the astronomical community to a new dimension, a gamma-ray universe where the energies are fantastic and the lifetimes are fleeting. While it is unlikely that gamma-ray astronomy will ever command the same attention as optical or x-ray astronomy, it has established itself as a discipline that all would-be or practicing astronomers should have some familiarity with.

This monograph attempts to bridge this cultural gap by summarizing the status of gamma-ray astronomy at energies above 30 MeV at a critical point in the development of the discipline: the hiatus between the demise of the Energetic

Gamma Ray Experiment Telescope (EGRET) telescope and the launch of the next generation space telescope, GLAST, as well as the hiatus before the completion of the next generation of imaging atmospheric Cherenkov detectors involving large arrays of telescopes. The present state of knowledge from observations of photons between 30 MeV and 50 TeV is summarized. Some attempt is made to describe the canonical explanations offered by theoretical models but this is still an observation-driven discipline. Although this branch of gamma-ray astronomy has been covered in previous works, this will be one of the first to focus on this energy band and to emphasize the higher energies.

Nothing dates a work more than a description of future developments but upcoming missions and projects are briefly described. In contrast, the early history is timeless and tells much. Each chapter has a brief historical note which describes a key development in that area. The principal processes by which gamma rays are produced and absorbed are well known and are well covered in standard physics texts. The appendix provides a brief summary of the most important processes.

Those who have worked in gamma-ray astronomy over the past four decades know what a wild and sometimes frustrating ride it has been. But I cannot think of a more exciting and exasperating profession nor can I imagine a more interesting time to be an astrophysicist in any discipline. That the discipline of gamma-ray astronomy has come to what it is today is in no small way due to the heroic efforts of those pioneers who more than 40 years ago gambled on there being a gamma-ray universe without even knowing there was an x-ray one. Those of us who followed those early pioneers have had the comfort of walking in their footprints and knowing that there was something to see at the end of the difficult path. Personally I have benefited greatly from the guidance of my early mentors, John Jelley and Neil Porter—physicists with their creativity and persistence are seldom encountered in my experience.

Gamma-ray astronomy has an artificial division at energies of about 100 GeV; below this energy the field thrives in the well-funded laboratories of space astronomy and above it, the work is done with more meagre resources by university groups using ground-based telescopes. Although the astrophysics of the sources does not recognize this energy break point, the two communities have a cultural divide and seldom overlap. In the intervals between operating gamma-ray satellites, the space community does not flock to use the ground-based instruments and equally the guest investigator programs of the space telescopes are not crowded with ground-based gamma-ray astronomers. This artificial divide is inevitably reflected in the subject matter of this monograph in which the two energies regions are often treated as if they were distinct.

In this work I have tried to emphasize the history as I know it. I have tried to be as accurate as possible but some things are a matter of interpretation. Inevitably there is some personal bias for which I have no apologies; it would be a sterile work if it did not reflect some personal opinions.

I am grateful to the many colleagues in the gamma-ray community who have shared their expertise and enthusiasm with me along the way; I am particularly grateful to members of the VERITAS gamma-ray collaboration who have been the stimulus for much of this work. Several colleagues read sections of the manuscript at various stages of production and made helpful suggestions; errors that remain are my responsibility. These readers included Mike Catanese, Valerie Connaughton, David Fegan, Stephen Fegan, Jerry Fishman, Jim Gaidos, Ken Gibbs, Michael Hillas, Deirdre Horan, Dick Lamb, Pat Moriarty, Simon Swordy and David Thompson. I am also appreciative of the many colleagues who supplied figures, including Michael Briggs, Werner Collmar, Stephen Fegan, Neil Gehrels, Alice Harding, Deirdre Horan, Stan Hunter, Kevin Hurley, John Kildea, Rene Ong, Toru Tanimori, and David Thompson. Irwin Shapiro has been a major supporter of VHE gamma-ray astronomy at the Smithsonian Astrophysical Observatory over the past two decades and I am proud to be a member of his staff. My wife, Ann, has, as always, been supportive and has also provided editorial assistance. I should also acknowledge the role of the funding agencies—but for their tardiness in funding the next generation of detectors I would have been hard put to find the time to put this work together.

Chapter 1

Foundations of gamma-ray astronomy

1.1 Astronomical exploration

Our knowledge of the physical universe beyond the earth comes almost entirely from the electromagnetic radiation received by our eyes or our manmade sensors. The environs of the earth, which we can explore directly, constitutes perhaps 10^{-58} times the volume of the universe. In our lifetime, mankind has seen the extension of the universe that can be physically explored with space probes to the distance of the Solar System's furthest planets. Human exploration thus far is limited to the moon, a tiny step on the cosmic scale. Although it is now feasible to consider unmanned space probes that will reach out to the nearest stars, it is still true that, in the foreseeable future, mankind will be limited to the observation of the radiations from distant sources as the sole means of exploring the distant cosmos.

It is important to emphasize that the astronomers who make a study of these radiations are always passive observers, never experimenters, in the sense that they do not control the experimental environment. This passive role is often a frustration to the high energy physicists who shift their interests into the realm of high energy astrophysics. The inability to control the experimental environment, to repeat the experiment to get better statistics, to vary the process with different input parameters... such limitations seem to make the astronomer powerless and a victim of circumstance.

But astronomers have two powerful weapons at their disposal: the number and variety of sources that they can observe; and the number of ways in which they can observe them. By observing a variety of versions of the same source, they can observe what they can hypothesize to be the same process, at different points in time. Moreover, by observing with the vast panoply of sensors now available, they can see the process in many different 'lights' and thence thoroughly explore the phenomenon. It is thus advantageous to use every conceivable band of the electromagnetic spectrum at its maximum sensitivity.

There is one other advantage that is uniquely available to them as astronomical observers: because they now have tools that permit the observation of sources at great distances they are also looking out at sources separated from them not only in distance but also in time. Thus they can consider the universe surrounding them to be like the layers of an onion; each layer is a chapter in the history of the universe and by comparing the differences in similar objects in adjacent layers they can see the evolution with time. The outermost layer is, of course, the beginning of time, the point when the expansion began and beyond which they have no knowledge. It is one of the outstanding contributions of modern astrophysics that we now have observations that pertain to the very first few seconds of this process. Modern cosmologists have become observational scientists but to continue their work they must use every tool at their disposal to probe these ultimate questions. Radiation that can penetrate great distances is thus of great value in these explorations.

It was inevitable that astronomers would want to explore every decade of the electromagnetic spectrum, no matter how far removed from ordinary terrestrial experience. Prior to the Second World War, the 'visible' band was the only really observational branch of astronomy but it was one that was extraordinarily rewarding since it was tuned to the peak in the spectrum of ordinary stars like our sun, to the transparency of the atmosphere, and to the sensitivity of the most accessible and versatile sensor, the human eye. The Second World War was to produce the radar technology that formed the basis of practical radio astronomy and the rocket technology that enabled x-ray astronomy. We can only speculate what the human perception of the cosmos would be if our human radiation sensors were in a band to which the atmosphere was largely opaque.

Photons are, by any definition, rather dull specimens in the cosmic particle zoo. However, one can argue that their very dullness, their lack of charge, mass, and moment, their infinite lifetime, their appearance as a decay product in many processes, their predictability, all combine to make them a valuable probe of the behavior of more exotic particles and their environs in distant, and therefore difficult to study, regions of the universe. Certainly no one can argue that photon astronomy at low energies (optical, radio and x-ray) has not largely shaped our perception of the physical universe!

1.2 The relativistic universe

Our universe is dominated by objects emitting radiation via thermal processes. The blackbody spectrum dominates, be it from the Big Bang (the cosmic microwave background), from the sun and stars, or from the accretion disks around neutron stars and other massive objects. This is the *ordinary* universe, in the sense that anything on an astronomical scale can be considered ordinary. It is tempting to think of the thermal universe as *THE UNIVERSE* and certainly it accounts for much of what we know about. However, to ignore the largely unseen, non-thermal, *extraordinary*, relativistic universe is to miss a major component and one that is of particular interest to the physicist, particularly the particle physicist. The relativistic universe is pervasive but largely unnoticed and involves physical processes that are difficult, if not impossible, to emulate in terrestrial laboratories. The most obvious local manifestation of this relativistic universe is the cosmic radiation, whose origin, 90 years after its discovery, is still largely a mystery (although it is generally accepted, *but not yet proven*, that much of it is produced in shock waves from galactic supernova explosions). The existence of this steady rain of relativistic particles, whose power-law spectrum confirms its non-thermal origin and whose highest energies extend far beyond that achievable in manmade particle accelerators, attests to the strength and reach of the forces that power this strange relativistic radiation. If thermal processes dominate the *ordinary* universe, then truly relativistic processes illuminate the *extraordinary* universe and must be studied, not just for their contribution to the universe as a whole but as the denizens of unique cosmic laboratories where physics is demonstrated under conditions to which we, terrestrial physicists, can only extrapolate.

The observation of the extraordinary universe is difficult, not least because it is masked by the dominant thermal foreground radiation. In some instances, we can see it directly such as in the relativistic jets emerging from active galactic nuclei (AGN) but, even there, we must subtract the overlying thermal radiation from the host elliptical galaxies. Polarization leads us to identify the processes that emit the radio, optical, and x-ray radiation as synchrotron emission from relativistic particles, probably electrons, but polarization is not unique to synchrotron radiation and the interpretation is not always unambiguous. The hard power-law spectrum of many of the non-thermal emissions immediately suggests the use of the highest radiation detectors to probe such processes. Hence, hard x-ray and gamma-ray astronomical techniques must play an increasingly prominent role among the observational disciplines of choice for the exploration of the relativistic universe.

The development of techniques whereby gamma rays of energy 100 GeV and above can be studied from the ground, using indirect, but sensitive, techniques is relatively new and has opened up a new area of high energy photon astronomy. The exciting results that have come from these studies include the detection of TeV photons from supernova remnants and from the relativistic jets in AGN.

Astronomy at energies up to a few GeV made dramatic progress with the launch of the Compton Gamma Ray Observatory (CGRO) in 1991. Beyond 10 GeV it is difficult to study gamma rays efficiently from space vehicles, both because of the sparse fluxes, which necessitate large collection areas, and the high energies, which make containment within a space telescope a serious problem.

The primary purpose of the astronomy of hard photons is the search for new sources, be they point-like, extended, or diffuse but this new astronomy also opens the door to the investigation of more obscure phenomena in extreme astrophysical environments and processes and even in cosmology and particle physics.

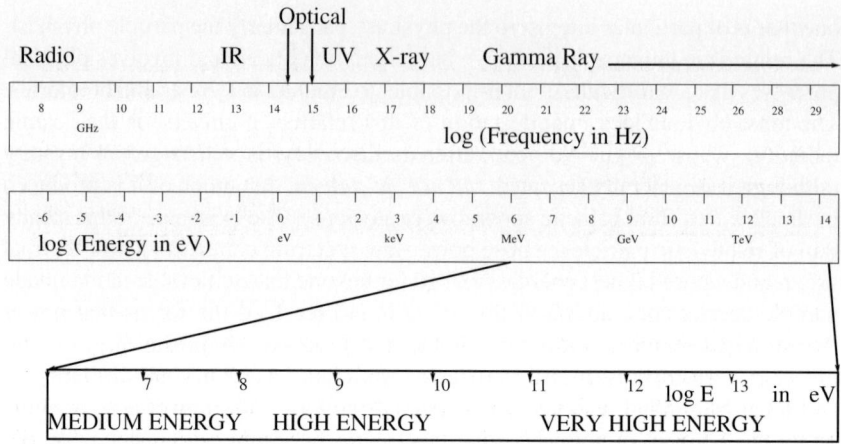

Figure 1.1. Electromagnetic spectrum showing the full extent of the part covered by the generic term, 'gamma rays'. The sub-divisions are defined in the text.

1.3 Definitions

The term 'gamma ray' is a generic one and is used to describe photons of energy from about 100 keV (10^5 eV) to >100 EeV (10^{20} eV). A range of 15 decades is more than all the rest of the known electromagnetic spectrum, i.e. from very long wavelength radio to hard x-rays (figure 1.1). A wide variety of detection techniques is, therefore, necessary to cover this huge band. This monograph will concentrate on the somewhat restricted gamma-ray band from 30 MeV to 100 TeV. The choice of this range is easy. It is the energy range where the detection techniques are relatively mature and have the maximum sensitivity; therefore, the best observational results have been obtained in these bands. Previous books [2, 12, 8, 10, 7, 11] have covered the full gamut of 'gamma-ray astronomy' above 100 keV with some loss of emphasis above 100 GeV where there were few results to report. There is, in fact, little in common between the phenomenon of nuclear line emission at MeV energies and the broad emission spectra of AGN at GeV–TeV energies. Hence it can be argued this restricted band of more than six decades (3×10^7 eV to 1×10^{14} eV) deserves a treatment on its own.

Even this band must be divided into two broad bands which are defined here, somewhat arbitrarily: the High Energy (HE) band from 30 MeV to 100 GeV and the Very High Energy (VHE) band from 100 GeV to 100 TeV (table 1.1). The band below 30 MeV (from about 1 to 30 MeV) is often called the Medium Energy (ME) region and that beyond 100 TeV, the Ultra High Energy (UHE) region. These gamma-ray regions are not defined by the physics of their production but by the interaction phenomena and techniques employed in their detection.

Table 1.1. Gamma-ray bands.

Band	Low/medium	High	Very High	Ultra High
Shorthand	LE/ME	HE	VHE	UHE
Range	0.1–30 MeV	30 MeV–100 GeV	100 GeV–100 TeV	>100 TeV
Typical energy	keV–MeV	MeV–GeV	TeV	PeV–EeV
Environment	Space	Space	Ground-based	Ground-based

Below 30 MeV, the Compton process is the dominant interaction process and Compton telescopes are used in their study; these techniques are difficult and inefficient but important because they include the potential study of nuclear lines. They will be only discussed briefly here. The detection techniques in the HE and VHE ranges use the pair-production interaction but in very different ways: HE telescopes identify the electron pair in balloon or satellite-borne detectors, whereas VHE detectors detect the resulting electromagnetic cascade that develops in the earth's atmosphere. As yet there are no credible detections of gamma rays at energies much beyond 50 TeV and hence the upper energy cutoff is a natural one at this time. Furthermore the 'gamma-ray telescope' techniques used beyond these energies are really the same as those used to study charged cosmic rays and, hence, are best studied in that context.

The boundaries of these bands are a matter of personal choice and different authors have defined the regions differently. However, most would agree that the HE region is characterized by observations in the 100 MeV range and the VHE region by observations around 1 TeV. That gamma-ray astronomy is still an observation-dominated discipline is apparent from these definitions.

1.4 The heroic era of gamma-ray astronomy

1.4.1 The early promise

Gamma rays are the highest energy photons in the electromagnetic spectrum and their detection presents unique challenges. On one hand, it is easy to detect gamma rays. The interaction cross sections are large and above a few MeV the pair production interaction, the dominant gamma-ray interaction with matter, is easily recognized. Gamma-ray detectors were already far advanced when the concept of 'gamma-ray astronomy' was first raised in Phillip Morrison's seminal paper in 1958 [9] (see historical note: seminal paper). Indeed it was the expected ease of detection and the early promise of strong sources that led to the large concentration of effort in this field, even before the development of x-ray astronomy. Today the number of known gamma-ray sources is well under a few hundred whereas there are hundreds of thousands of x-ray sources. Why have the two fields developed so differently?

The answer is simple: the detection of cosmic gamma rays was not as easy as expected and the early predictions of fluxes from cosmic sources were hopelessly optimistic.

1.4.2 Peculiarities of gamma-ray telescopes

There are several peculiarities that uniquely pertain to astronomy in the gamma-ray energy regime. These factors make gamma-ray astronomy particularly difficult and have resulted in the relatively slow development of the discipline.

In nearly every band of the electromagnetic spectrum, astronomical telescopes make use of the fact that the cosmic rain of photons can be concentrated by reflection or refraction, so that the dimensions of the actual photon detector are a small fraction of the telescope aperture. How limited would have been our early knowledge of the universe if the optical astronomer had not been aided by the simple refracting telescope which so increased the sensitivity of the human eye! The radio astronomer, the infrared astronomer, even the x-ray astronomer, depends on the ability of a solid surface to reflect and, with suitable geometry, to concentrate the photon signal so that it can be detected above the background by a small detector element.

Above a few MeV, there is no efficient way of reflecting gamma rays and hence the dimensions of the gamma-ray *detector* are effectively the dimensions of the gamma-ray telescope. (As we shall see in the next chapter this is not the case for ground-based VHE telescopes.) In practice, to identify the gamma-ray events from the charged particle background it is necessary to use detectors whose efficiency is often quite low. Hence, at any energy the effective aperture of a space-borne gamma-ray telescope is seldom greater than 1 m^2 and often only a few cm^2, even though the physical size is much larger. The Compton Gamma Ray Observatory was one of the largest and heaviest scientific satellites ever launched; however, its ME and HE telescopes had effective apertures of 5 cm^2 and 1600 cm^2 respectively. Beam concentration is particularly important when the background scales with detector area. This is always the case with gamma-ray detectors which must operate in an environment dominated by charged cosmic rays.

The problem of a small aperture is compounded by the fact that the flux of cosmic gamma rays is always small. At energies of 100 MeV the strongest source (the Vela pulsar) gives a flux of only one photon per minute in telescopes flown to date. With weaker sources, long exposures are necessary and one is still dealing with the statistics of small numbers. Small wonder that gamma-ray astronomers have been frequent pioneers in the development of statistical methods and that early gamma-ray conferences were often dominated by arguments over real statistical significances! As it is to photons in many bands of the electromagnetic spectrum, the earth's atmosphere is opaque to all gamma rays. Even the highest mountain is many radiation lengths below the top of the atmosphere so that it is virtually impossible to consider the direct detection of cosmic gamma rays without the use of a space platform. Large balloons can carry the bulky detectors

Figure 1.2. The Lebedev Institute experiment that operated in the Crimea, c. 1960–64. This was the first major VHE gamma-ray telescope. (Photo: N A Porter.)

to near the top of the atmosphere and much of the pioneering work in the field was done in this way. However, the charged cosmic rays constitute a significant background and limit the sensitivity of such measurements.

The background can take many forms. In deep space it is the primary cosmic radiation itself, mostly protons, heavier nuclei and electrons. This background can be accentuated by secondary interactions in the spacecraft. Careful design and shielding can reduce this effect, as can active anti-coincidence charged-particle shields. However, at low energies induced radioactivity in the detector and its surrounds can be a serious problem. In balloon experiments gamma rays in the secondary cosmic radiation from the cosmic ray interactions in the atmosphere above the detector seriously limit the sensitivity and were the initial reason for the slow development of the field. Huge balloons that carry the telescopes to within a few grams of residual atmosphere are a partial solution but it is still impossible to trust the measurement of absolute diffuse fluxes.

1.4.3 VHE gamma-ray telescopes on the ground

Shortly after the detection of atmospheric Cherenkov radiation (see appendix) from cosmic ray air showers, the phenomenon was utilized to look for point-

8   Foundations of gamma-ray astronomy

Figure 1.3. The Whipple 10 m gamma-ray telescope. Note the '10 m' refers only to the aperture of the optical reflector; the effective collection area is $>5 \times 10\,000$ m^2 so that the gamma-ray 'aperture' is 120 m.

source anomalies in the cosmic ray arrival direction distribution which might point to the existence of discrete sources of VHE cosmic rays. None were found. Not long after the publication of Morrison's seminal paper [9] on the prospects for gamma-ray astronomy at 100 MeV energies (see historical note: seminal paper), Cocconi, a high energy theorist at CERN, produced an equally optimistic prediction for the possibilities of gamma-ray astronomy at VHE energies [5]. He made his predictions for telescopes consisting of arrays of particle detectors. Two such experiments (in Poland and Bolivia) searched for discrete sources but their energy thresholds were high (>100 TeV) and no anomalies were found. Other experimenters realized that the detection of the electromagnetic cascades using the atmospheric Cherenkov radiation was a more sensitive technique and an ambitious array of 12 light detectors was deployed in the Crimea by a group from the Lebedev Institute (figure 1.2). Four years of operation (1960–64) by the Soviet group [3] produced extensive observations of the sources suggested by Cocconi (radio galaxies and supernova remnants) but did not lead to any source

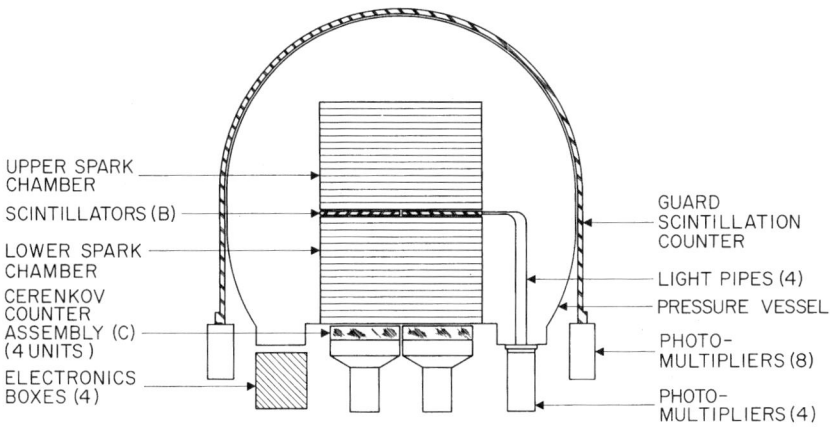

Figure 1.4. The pair production spark chamber telescope on the SAS-2 satellite [6]. (Figure: D Thompson.)

detections. Nevertheless, they laid the groundwork for the future development of the technique.

In the years that followed, more than a dozen ground-based experiments sought to extend these observations and to improve the techniques. The first large optical reflector purpose-built for gamma-ray astronomy was the Whipple Observatory's 10 m optical reflector (installed on Mount Hopkins in southern Arizona in 1968) (figure 1.3). Although this larger collection area led to a reduction in energy threshold, it did not immediately lead to a significant improvement in flux sensitivity. The apparent detection of a signal from the x-ray binary, Cygnus X-3, by groups in the Soviet Union, in Germany, and in the United Kingdom, using both atmospheric Cherenkov techniques and air shower particle arrays, led to an upsurge in experimental activity but no major improvements in detection technique. There were unsubstantiated claims for the detection of signals from a variety of binaries and pulsars but the signals were transient and of marginal statistical significance.

The subsequent development of the atmospheric Cherenkov imaging technique, using the Whipple telescope 20 years later, led to the detection of the Crab Nebula in 1989 [13]. This detection of a steady source, which has become the 'standard candle' for the field, ended this period of uncertainty in the development of VHE gamma-ray astronomy.

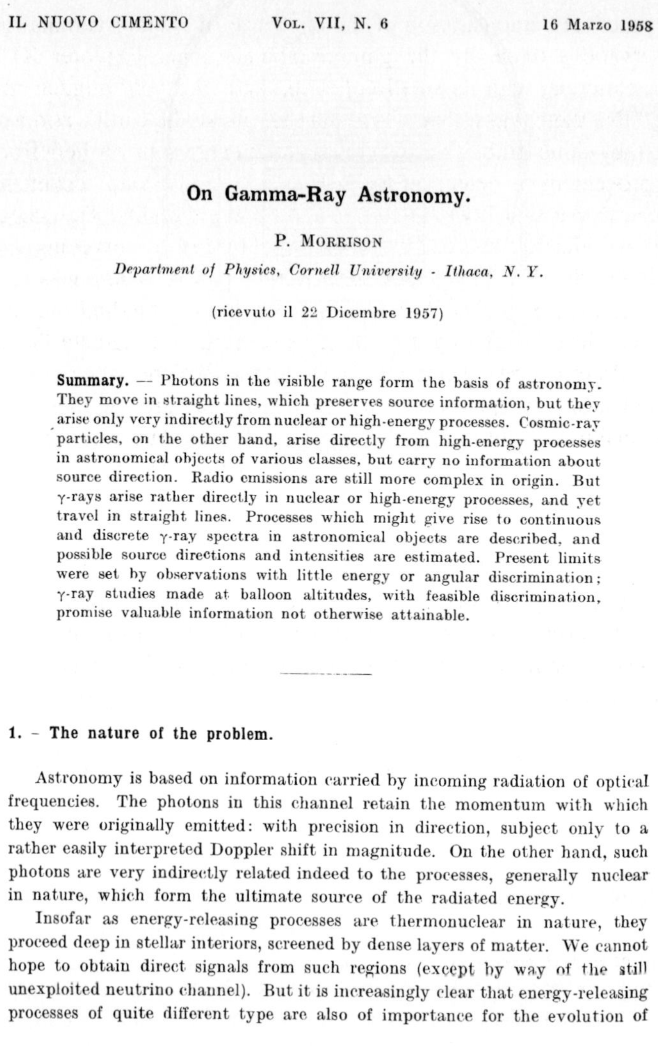

IL NUOVO CIMENTO   Vol. VII, N. 6   16 Marzo 1958

On Gamma-Ray Astronomy.

P. MORRISON

Department of Physics, Cornell University - Ithaca, N.Y.

(ricevuto il 22 Dicembre 1957)

Summary. — Photons in the visible range form the basis of astronomy. They move in straight lines, which preserves source information, but they arise only very indirectly from nuclear or high-energy processes. Cosmic-ray particles, on the other hand, arise directly from high-energy processes in astronomical objects of various classes, but carry no information about source direction. Radio emissions are still more complex in origin. But γ-rays arise rather directly in nuclear or high-energy processes, and yet travel in straight lines. Processes which might give rise to continuous and discrete γ-ray spectra in astronomical objects are described, and possible source directions and intensities are estimated. Present limits were set by observations with little energy or angular discrimination; γ-ray studies made at balloon altitudes, with feasible discrimination, promise valuable information not otherwise attainable.

---

1. – The nature of the problem.

Astronomy is based on information carried by incoming radiation of optical frequencies. The photons in this channel retain the momentum with which they were originally emitted: with precision in direction, subject only to a rather easily interpreted Doppler shift in magnitude. On the other hand, such photons are very indirectly related indeed to the processes, generally nuclear in nature, which form the ultimate source of the radiated energy.

Insofar as energy-releasing processes are thermonuclear in nature, they proceed deep in stellar interiors, screened by dense layers of matter. We cannot hope to obtain direct signals from such regions (except by way of the still unexploited neutrino channel). But it is increasingly clear that energy-releasing processes of quite different type are also of importance for the evolution of

Figure 1.5. The seminal paper in *Il Nouvo Cimento* on gamma-ray astronomy [9]; this journal was a traditional location for papers on cosmic ray studies.

Historical note: seminal paper

Reproduction of the first page of the paper by Morrison [9] (figure 1.5) which is usually credited as being the seminal paper for gamma-ray astronomy (with permission from *Il Nuovo Cimento*).

1.4.4 HE gamma-ray telescopes in space

The first practical demonstration of the existence of cosmic gamma rays came from observations made by the gamma-ray telescope, Explorer XI in 1965 [4]. This telescope, with its small collection area and poor angular resolution, established that there was a flux of gamma rays above the earth's atmosphere but not where they came from. This result was sufficient to spur further efforts aimed at the improvement of detection techniques. Several groups developed spark chamber telescopes which were flown in short balloon flights. The objective was the detection of discrete sources at energies of 100 MeV by surveying a relatively small region of the sky. This was a controversial period. There was no gamma-ray source equivalent of Sco X-1 in x-ray astronomy, standing out like a sore thumb, to confirm the existence of discrete sources and validate the detection techniques. As might be expected in a new field with pioneering detectors, weak sources, and strong competition between experimental groups, there were many conflicting claims on source detection but all had weak statistics. With hindsight, the detection of the Crab pulsar was the first credible detection [1] and served to motivate the field to develop new techniques.

The balloon era of HE gamma-ray exploration came to an effective end with the launch of NASA mission SAS-2 in 1972 [6] (figure 1.4). This was the prototype spark chamber gamma-ray telescope and although it had an unexpectedly short lifetime (six months) because of a power supply failure, it laid the groundwork for all future gamma-ray space missions. The galactic plane was mapped, discrete sources were discovered and the diffuse background established. These results were confirmed and greatly extended by the European gamma-ray satellite, COS-B, which was launched in 1975 and which enjoyed a lifetime of seven years.

References

[1] Browning R, Ramsden D and Wright P J 1971 *Nature* **232** 99
[2] Chubb E L 1976 *Gamma Ray Astronomy* (Dordrecht: Reidel)
[3] Chudakov A E, Dadykin V I, Zatsepin and Nestrova N M 1965 *Transl. Consultants Bureau, P. N. Lebedev Phys. Inst.* **26** 99
[4] Clark G W, Garmire G P and Kraushaar W L 1968 *Astrophys. J. Lett.* **153** L203
[5] Cocconi G 1959 *Proc. Int. Cosmic Ray Conf. (Moscow)* **2** 309
[6] Derdeyn S M *et al* 1972 *Nucl. Instrum. Methods* A **98** 557
[7] Fichtel C E and Trombka J I 1997 *Gamma-Ray Astrophysics: New Insight Into the Universe (NASA Reference Publication 1386)* 2nd edn
[8] Hillier R 1984 *Gamma Ray Astronomy* (Oxford: Oxford University Press)
[9] Morrison P 1958 *Nuovo Cimento* **7** 858
[10] Ramana Murthy P V and Wolfendale A W 1993 *Gamma Ray Astronomy* (Cambridge: Cambridge University Press)
[11] Schonfelder V 2001 *The Universe in Gamma Rays* (Berlin: Springer)

[12] Stecker F W 1971 *Cosmic Gamma Rays (Publ. NASA SP-249)* (Baltimore, MD: Mono Book Corporation)
[13] Weekes T C *et al* 1989 *Astrophys. J.* **342** 379

Chapter 2

Very high energy gamma-ray detectors

2.1 The atmospheric windows

It is natural that astronomy should develop using those parts of the electromagnetic spectrum to which the atmosphere is transparent and for which detectors were available. This led to the early development of optical astronomy (at the dawn of mankind) and radio astronomy as radio and radar techniques were developed in the period around the Second World War. The earth's atmosphere effectively blocks all electromagnetic radiation of energies greater than 10 eV. The total vertical thickness of the atmosphere above sea level is 1030 g cm^{-2} and since the radiation length is 37.1 g cm^{-2}, this amounts to more than 28 radiation lengths. This is equivalent in blocking power to a 1 m thickness of lead. This is true up to the energy of the highest known cosmic rays (some of which may be gamma rays). Much of the electromagnetic spectrum was not available until space techniques, first rockets and balloons, and later satellites, became accessible. Hence, until 1960, almost all astronomical observations came via the radio and optical windows. It may seem nonsensical then to speak of a 'gamma-ray window' where ground-based telescopes can make observations since no significant flux of primary gamma rays can penetrate even to the elevation of the highest mountain. However, there is a 'gamma-ray window' from about 100 GeV to 50 TeV where it has been possible to successfully pursue gamma-ray observations of cosmic sources using ground-based instruments. It is a fortunate coincidence in nature that while the gamma ray itself may not survive, the secondary products of its interaction with the atmosphere do survive and can be detected with the simple detectors described here. The techniques that are used in this window are described in this chapter. It is also a coincidence that the minimum energy that the gamma ray must have to be detectable from the ground is just above the maximum energy that has been detected by the space telescopes described in the following chapter.

2.2 Electromagnetic cascade in atmosphere

The predominant interaction of a gamma ray of energy greater than 10 MeV, as it enters the earth's atmosphere, is pair production (see appendix). Typically this will occur after it traverses one radiation length of atmosphere, i.e. at an altitude of about 20 km. The resultant electron–positron pair will share the energy of the primary gamma ray and will be emitted in the forward direction [6]. Hence, for gamma rays of energy 10 GeV or larger, the gamma ray is effectively replaced by two charged particles travelling in almost the same direction as the gamma ray.

After they have traversed a radiation length, on average, these particles will interact with air molecules to give secondary gamma rays by the bremsstrahlung process. After another radiation length these secondary gamma rays may also pair produce. The angle of emission in all these processes will be $\propto m_e c/E$ rad, where E is the energy of the electron and m_e is the rest mass of the electron. The resulting electromagnetic cascade will be remarkably tightly bunched along the projection of the original gamma-ray trajectory.

The process continues down through the atmosphere with the number of secondary electrons, positrons and gamma rays increasing until the average energy drops to a point where ionization energy losses and the radiation losses become equal. At this point the cascade reaches 'shower maximum' (N_{max} = maximum number of electrons, h_{max} = the elevation at which this occurs in km, and X_{max} is the shower thickness traversed in g cm^{-2}). The number of particles gradually diminishes and the cascade dies away. Depending on the primary gamma-ray energy, this may be well before it reaches sea level. N_{sl} = number of surviving particles at sea level and N_{mt} = number at mountain altitude (2.3 km).

For typical gamma-ray primaries the value of N_{max}, N_{sl}, N_{mt}, X_{max} and h_{max} are tabulated in table 2.1. If the secondary electrons are above the threshold for Cherenkov emission (see appendix), they will cause the atmosphere to radiate in the forward direction (Cherenkov angle $\theta \approx 1.3°$ at sea level where the refractive index, $n = 1.00029$) [13]. The development of an electromagnetic cascade is shown schematically in figure 2.1, where the trajectory of each particle above the Cherenkov threshold is calculated in a Monte Carlo simulation. Since many of the electrons and positrons will be above the threshold (21 MeV at sea level), the electromagnetic cascade will be accompanied by a shower of Cherenkov photons which will suffer little atmospheric absorption and whose density at sea level within 120 m of the shower axis can be characterized by an optical photon density (300–450 nm) of ρ_{sl} photons m^{-2}. The photon density at mountain altitude, ρ_{mt} (photons m^{-2}), is also tabulated in table 2.1. These values come from Monte Carlo calculations by A M Hillas [11].

2.3 The visible electromagnetic cascade

It is convenient to think of the shower as a glowing column of light as seen by an observer on the ground. In many ways (despite the beaming effect of the

Table 2.1. Gamma-ray shower parameters as a function of energy [11].

Energy, E_γ	X_{max} (g cm^{-2})	h_{max} (km)	N_{max}	N_{sl}	N_{mt}	ρ_{sl} (photon m^{-2})	ρ_{mt} (photon m^{-2})
10 GeV	175	12.8	1.6×10^1	4×10^{-4}	2×10^{-2}	2.7×10^{-1}	3.6×10^{-1}
100 GeV	261	10.3	1.3×10^2	4.0×10^{-2}	1.4×10^0	4.6×10^0	7.6×10^0
1 TeV	346	8.4	1.1×10^3	3×10^0	6.0×10^1	7.4×10^1	1.3×10^2
10 TeV	431	6.8	1.0×10^4	1.3×10^2	1.7×10^3	1.1×10^3	1.7×10^3
100 TeV	517	5.5	9.3×10^4	4.5×10^3	3.6×10^4	1.6×10^4	1.9×10^4
1 PeV	602	4.4	8.6×10^5	1.15×10^5	5.7×10^5	1.9×10^5	1.9×10^5

16   *Very high energy gamma-ray detectors*

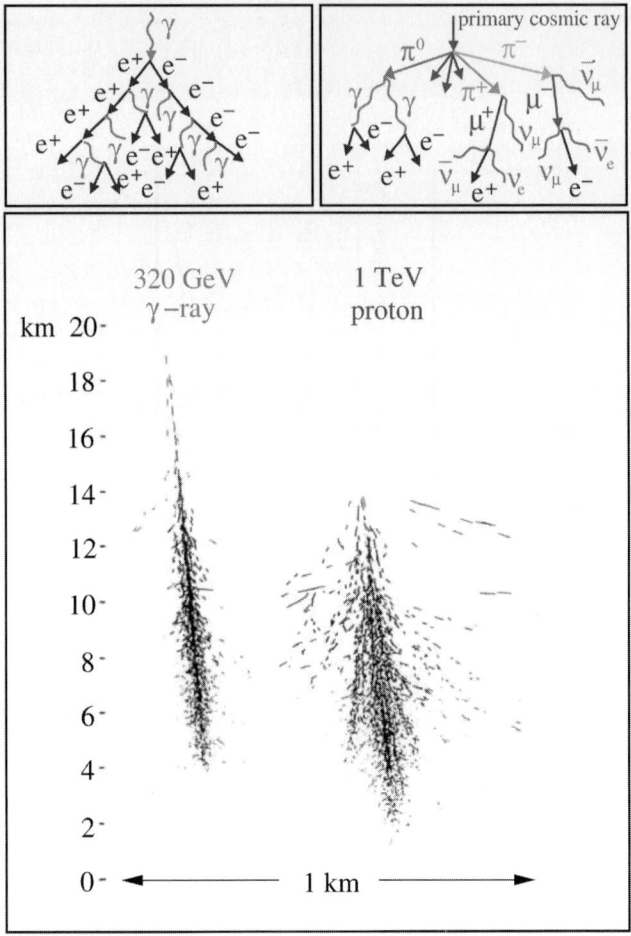

Figure 2.1. Main panel: Monte Carlo simulations of 320 GeV gamma-ray shower and a 1 TeV proton shower. The tracks of Cherenkov light emitting particles are shown but not all to avoid saturation. The horizontal scale is magnified by a factor of five [10]. A schematic development of a gamma-ray shower (left) and a hadronic shower (right) are shown in the two small panels. (Figure: D Horan.)

Cherenkov emission), it is similar to the trail of a meteor. In particular, the column seen on the sky when extrapolated backwards intersects the point of origin of the gamma ray on the cosmic sphere. The optical images of a shower of meteors have a similar property in that they all point back to their point of origin, i.e. the radiant of the meteor shower.

Although the fraction of energy that goes into this optical emission is small (less than 10^{-6} of the primary energy), it is coherent and this makes possible a very simple method for detecting the cascade and, thence, the gamma ray. A simple light detector (mirror plus phototube plus fast pulse counting electronics) provides an easy way of detecting the cascade.

The astronomer is interested in detecting the gamma ray and then determining its point of origin (so that a map of sources on the celestial sphere can be constructed), its energy (so that the energy spectrum of the sources can be determined), and its time of arrival (so that variability in the source emission can be determined). At the energies of interest (>100 GeV), the shower core develops along the projected trajectory of the primary to a high degree. Since Cherenkov light is emitted from all the particles in the shower whose energy is above the Cherenkov threshold, the atmosphere behaves like a giant calorimeter and, hence, the measurement of the brightness of the light is a good measure of the energy of the primary gamma ray. The Cherenkov light arrives at detector level within a span of a few nanoseconds, so the time of arrival of the gamma ray can be recorded to high precision.

Gamma-ray studies in the 1 TeV energy range are greatly facilitated by the fact that the physics of the principal particle interactions at these energies is well understood and modern computers can be used to simulate individual showers with high accuracy. The average properties of the electromagnetic cascades and their Cherenkov light emission can be simulated using Monte Carlo techniques (figure 2.1). Such simulations have been found to agree well with the measured properties of gamma-ray-initiated air showers.

The Cherenkov light, as seen at detector level, can be considered to come from the three portions of a typical 1 TeV shower shown schematically in figure 2.2 [10]. The first portion (containing ∼25% of the total light) comes from shower particles at elevations between the height of the first interaction down to an elevation of 10 km. The Cherenkov angle broadens with decreasing altitude with the net effect that the light appears as a 'focused' annulus on the ground of radius about 120 m. The light from the highest altitude arrives at the same time as that from the lower altitudes; the particles in the shower travel close to the velocity of light and compensate for the greater geometric path travelled because light from the higher altitudes goes at the slower velocity of c/n in air (see appendix). The result is that the light in the annulus is strongly bunched in time with a spread of ∼1 ns.

The bulk of the light (∼50%) comes from a cylinder of length 4 km and radius 21 m centered on the shower core; this cylinder contains the shower maximum and, hence, the bulk of the light emitted. The light from this region is a good measure of the total energy, i.e. this cylinder is the best calorimeter and the light is best measured at a distance of ∼100 m from the shower axis on the ground. The angular spread of this light will have a half-width of ∼0.2°.

The last 25% of the light comes from the local component of the shower, particles radiating below an elevation of 6 km. This light generally falls close to the shower axis intersection point and is subject to large fluctuations because it is dominated by the few surviving particles.

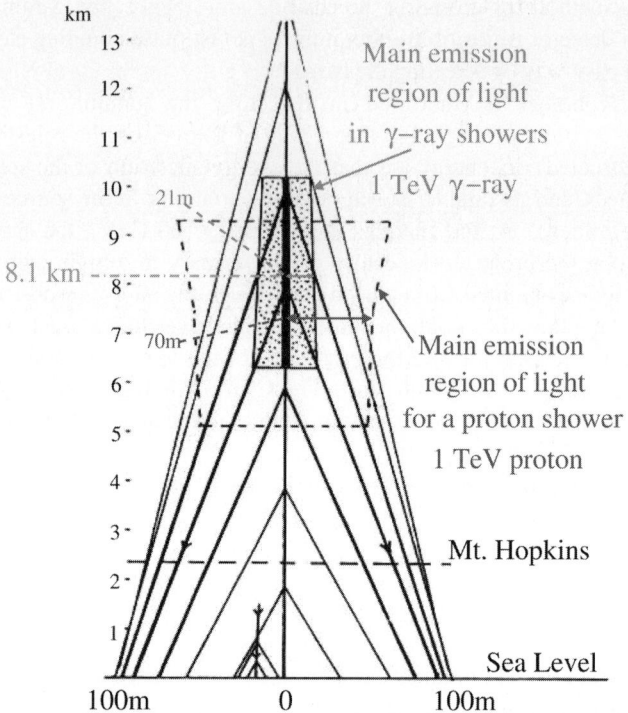

Figure 2.2. Cartoon showing the Cherenkov light emitting regions of gamma-ray and proton air showers. The shaped area corresponds to the main region of emission in a gamma-ray shower of 1 TeV energy. The area enclosed by the broken line is the main region of emission for 1 TeV proton shower. The lateral distribution of light from the gamma-ray shower is shown at the bottom of the diagram. Note that the horizontal scale is magnified by a factor of five [10]. (Figure: A M Hillas.)

2.4 Atmospheric Cherenkov technique

2.4.1 General properties

The basic atmospheric Cherenkov telescope (ACT) can be very simple [25, 4, 19]. First-generation systems consisted of just a single light detector in the focal plane of searchlight mirror coupled to fast pulse counting electronics. The basic elements are illustrated in figure 2.3. Such telescopes are characterized by the mirror collection area, A, the reflectivity, R, the solid angle, Ω, and the integration time, τ. Even with a simple light detector ($A = 2$ m^2, $R = 85\%$, $\Omega = 10^{-3}$, and $\tau = 10$ ns), it is possible to detect the light signal from gamma-ray showers of a few TeV energy with high efficiency. Its identification as coming from a gamma-ray shower rather than from a cosmic ray air shower is quite a different matter

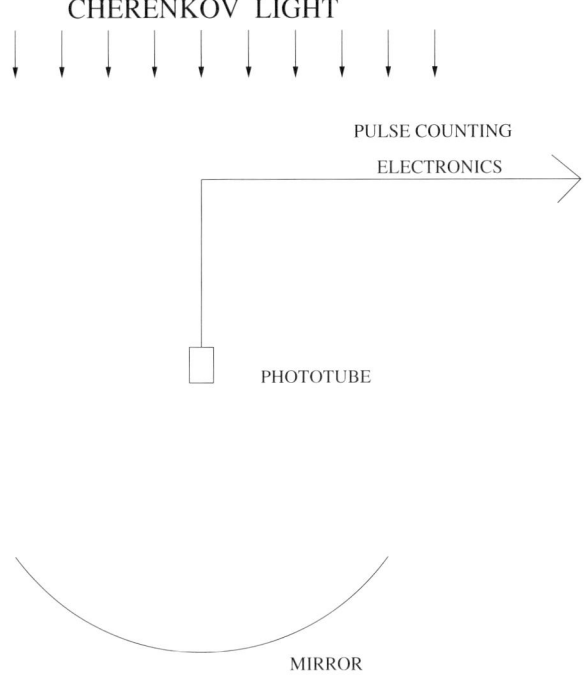

Figure 2.3. Schematic diagram of simple atmospheric Cherenkov gamma-ray telescope.

(see later). A bias curve (counting rate pulse height) (figure 2.4) taken with this simple detector shows two components: a soft (steeply falling) component with a power-law exponent ~ -7 and a hard component with a power exponent of -1.7. The soft component is due to fluctuations in the night-sky background light; the hard component comes from the Cherenkov light flashes from air showers (mostly initiated by hadrons). Hence, the exponent of this component is approximately that of the primary cosmic ray spectrum.

2.4.1.1 Signal

If the integration time of the photomultiplier pulse counting system, τ, is greater than the duration of the Cherenkov light flash (typically 3–5 ns), then the light signal (in photoelectrons) detected is given by

$$S = \int_{\lambda_2}^{\lambda_1} C(\lambda)\eta(\lambda)A\,\mathrm{d}\lambda$$

where $C(\lambda)$ is the Cherenkov photon flux within the wavelength sensitivity bounds of the PMT, λ_1 and λ_2, and $\eta(\lambda)$ is the response curve of the PMT.

$$C(\lambda) = kE(\lambda)T(\lambda)$$

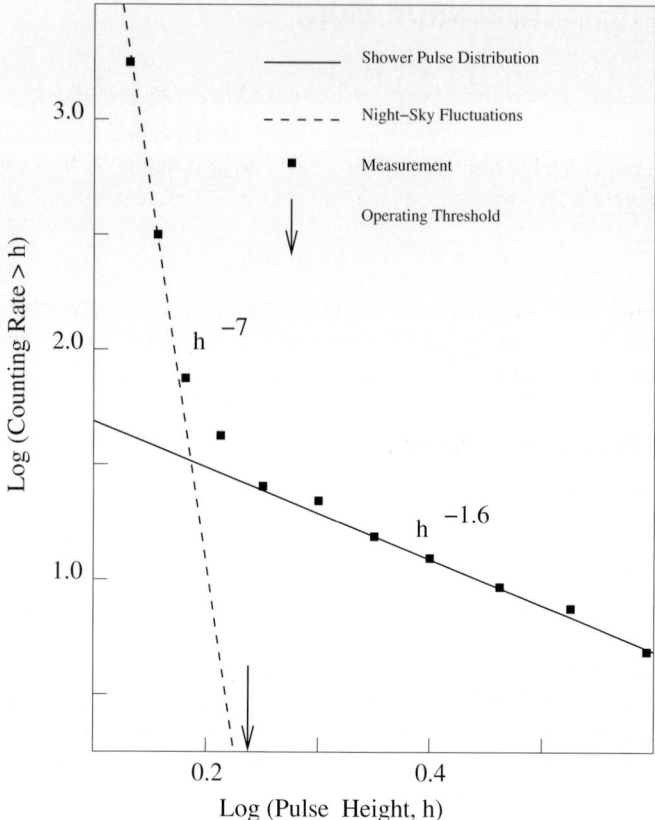

Figure 2.4. The pulse height distribution seen when the telescope shown in figure 1.4 is exposed to the night-sky. The arrow indicates the typical operating threshold where the detector is seldom triggered by night-sky fluctuations.

where $E(\lambda)$ is the shower Cherenkov emission spectrum ($\propto 1/\lambda^2$), $T(\lambda)$ is the atmospheric transmission (figure 2.5) and k is a constant which depends on the number of particles in the shower, and the geometry of the emitting particles and detector.

2.4.1.2 Background

The Cherenkov light pulse must be detected above the fluctuations in the night-sky background in the time interval, τ.

The sky noise B is given by

$$B = \int_{\lambda_2}^{\lambda_1} B(\lambda)\eta(\lambda)\tau A\Omega \, d\lambda.$$

Hence the signal-to-noise ratio is essentially

$$S/N = S/B^{0.5} = \int_{\lambda_2}^{\lambda_1} C(\lambda)[\eta(\lambda)A/\Omega B(\lambda)\tau]^{1/2}\,d\lambda.$$

The smallest detectable light pulse is inversely proportional to S/N; the minimum detectable gamma ray then has an energy threshold, E_T given by

$$E_T \propto 1/C(\lambda)[B(\lambda)\Omega\tau/\eta(\lambda)A]^{1/2}.$$

For the signal to be identified as coming from other than an extreme fluctuation in the ambient light background, it must be ∼5–7 times N, depending on the configuration of the detector electronics.

2.4.2 Features of the technique

2.4.2.1 Atmosphere

To most astronomers, the earth's atmosphere is a troublesome filter which distorts and limits their observations. There are only a few wavebands in the electromagnetic spectrum to which the atmosphere is transparent or partially transparent. Even at visible wavelengths, where it is remarkably transparent, the turbulence distorts the images and, ultimately, limits the angular resolution. It is quite different for the VHE gamma-ray astronomer; the atmosphere is an essential ingredient in the detection technique, a free, and almost limitless, component that makes detection possible. However, as with many bargains, it exacts a price in other ways. It is not a component over which the gamma-ray astronomer has any control; it varies in temperature, pressure, and humidity and, thus, changes the characteristics of the telescope. More troublesome, the atmospheric extinction changes so that transmission is a variable. The presence of thin cirrus clouds is always difficult to detect, although most of the optical emission comes lower in the atmosphere. The atmospheric parameters can be monitored and are carefully measured in the new generation of telescopes. It is remarkable that many of the results achieved to date have not had the benefit of such monitoring, indicating that these atmospheric parameters are second-order effects.

2.4.2.2 Light collectors

To maximize the sensitivity to Cherenkov light detection, the light collection area of an atmospheric Cherenkov telescope must be as large as possible. The angular size of the image of a Cherenkov light shower is 0.5–1.0° and the meaningful structure a few arc-min. Hence, optical collectors (telescopes) for ground-based gamma-ray detection do not have to approach the standards of optical astronomy telescopes and can be constructed relatively inexpensively. The favorite method of achieving large areas at low cost is the use of tessellated arrays of spherical mirrors of the same focal length. If these are located on an optical support structure with the same radius of curvature as the focal length (Davis–Cotton design), then the optimum optical image is achieved. This design gives a good optical image within a few degrees of the optic axis; however, it introduces a time spread in the time of arrival of the light in the focal plane. In the Whipple $f/0.7$ 10 m reflector (figure 1.3), this spread is about 6 ns; it is less in collectors with larger f-numbers.

The individual mirror segments are usually made of glass, round or hexagonal in shape (for close packing), front-aluminized (to give good ultraviolet response), and of diameter 60–100 cm (to permit easy handling). Because of the large overall size of the light collector, it is usually not protected by a dome or cover. Hence, weathering of mirror surfaces is a problem. For this reason, the aluminum surfaces are usually anodized. The mirrors must be regularly cleaned. The largest aperture telescope currently in use is that at the Whipple Observatory; this was built in 1968 and has an aperture of 10 m. The MAGIC telescope will soon come into use with an aperture of 17 m [16]. In principle, the mirror collection area can be increased until it is of the same order as the dimensions of the shower light pool (radius $\sim$120 m).

2.4.2.3 Light detectors

Fast, blue-sensitive, broadband light detectors are available in the form of photomultipliers (PMTs). Because of their many uses, these light detectors are readily available at reasonable cost. The peak quantum efficiency is typically 15% and the response curve $\eta(\lambda)$ as a function of wavelength, λ, has the form shown in figure 2.5. The disadvantages of PMTs are that they operate at high voltage and can be easily damaged by excessive light. Nonetheless, these detectors have been the workhorses of all ACT systems to date. There is much interest in the development of a blue-sensitive detector of greater quantum efficiency, probably a hybrid solid-state photomultiplier or avalanche photodiode. In principle, this is a less expensive way of reducing the energy threshold than increasing the mirror aperture.

2.4.2.4 Sky brightness

The duration of the Cherenkov light pulse is $\sim$3–4 ns and the detector response must be matched to this short duration. Ultimately, the energy threshold for gamma-ray detection is determined by the background light, which must be minimized for maximum sensitivity. Potential sources of background light are represented in the cartoon in figure 2.6. Although the atmosphere comes at no cost, the observer has no control over it; the telescope is wide open to the elements and the detector is susceptible to a troublesome background of light from sun, moon and stars, from airglow, from lightning and meteors, and from a variety of manmade light sources, e.g. from satellites, airplanes, beacons, and city lights. These light sources limit the sensitivity for gamma-ray source detection. Since

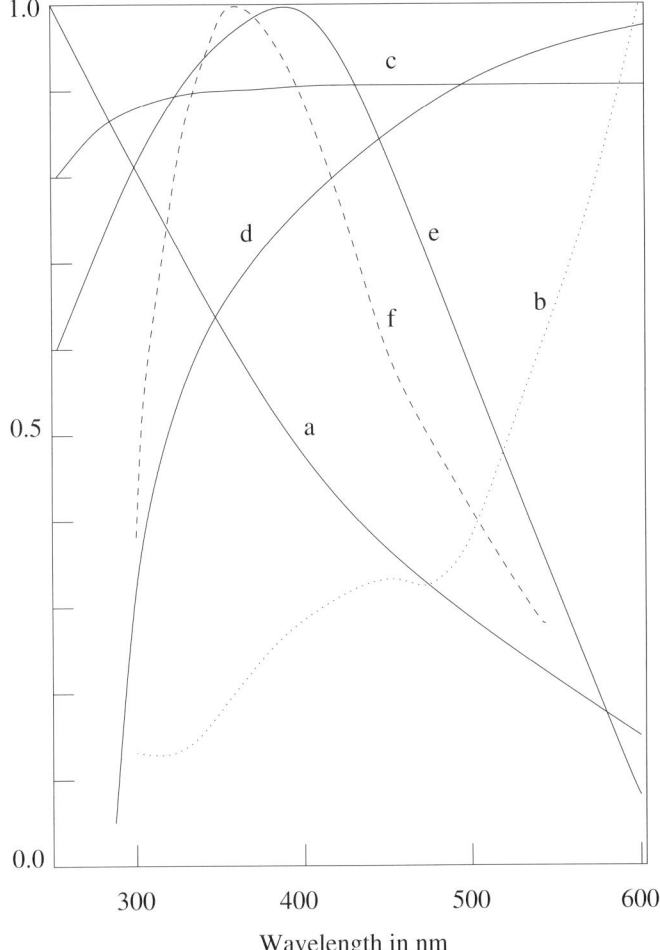

Figure 2.5. (a) The variation of Cherenkov light emission yield relative to emission at 250 nm. (b) Spectral distribution of the night-sky background relative to that at 600 nm (atomic lines omitted). (c) Reflectivity of typical mirror. (d) Transmission through atmosphere from 10 to 2.3 km. (e) Quantum efficiency of bi-alkali photocathode relative to that at 385 nm. Product of (a), (c), (d) and (e) normalized to 1.0 at 350 nm. All quantities are plotted as a function of wavelength.

one is detecting short pulses of light, variable sources, e.g. airport beacons, are more detrimental than steady sources, e.g. city lights. By choice of site away from manmade lights, the background light can be minimized. By choice of observing time one can avoid the sun, moon, and lightning. It is more difficult to minimize the natural background due to starlight and airglow. These have a

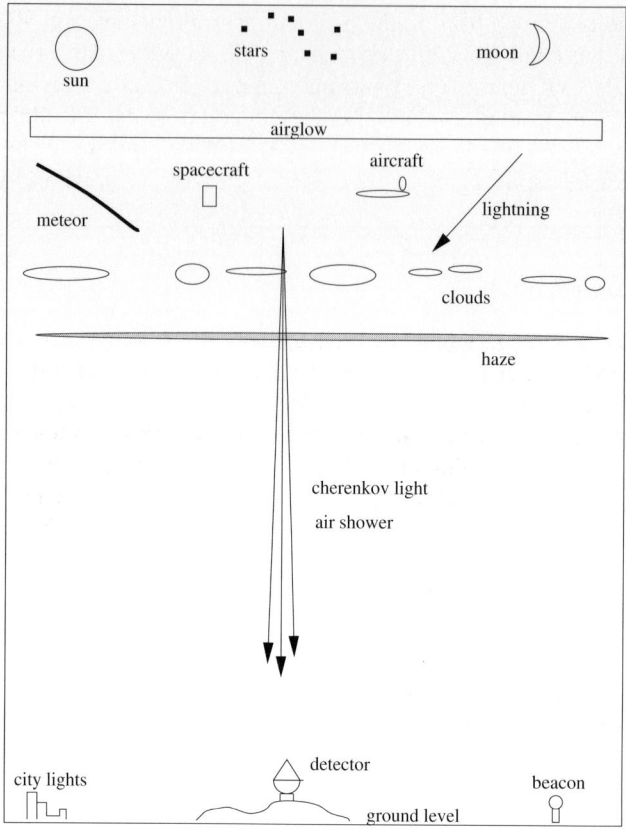

Figure 2.6. Sources of background light for atmospheric Cherenkov telescopes showing both natural and manmade sources. The sources that are time variable are the most troublesome.

broadband emission spectrum, as has Cherenkov light emission. The latter is peaked towards the ultraviolet, so it is best to choose a photodetector that has high quantum efficiency in the blue, consistent with atmospheric transmission. Generally, it has not been found to be advantageous to use filters to optimize the signal-to-background. Although a filter can be found that will preferentially transmit Cherenkov light, in practice the reduced transmission nullifies the gain and Cherenkov detectors invariably define their waveband by the response of the photodetector.

The Cherenkov light, given off by the passage of the electromagnetic cascade through the atmosphere, can only be detected if it is brighter than the background light from other sources. If a dark site is chosen, away from manmade lights, then the optimum observing conditions are during clear moonless nights. The optimum optical band is a broad band in the blue and near ultraviolet, e.g. 300–450 nm. Fortunately, this band is one in which sensitive PMTs are available (with bi-alkali photocathodes and quartz or uv transmitting glass). It is also a region where the background light from the night sky is near minimum (in the 300–400 nm band it is about 2.5×10^{-4} erg s^{-1} cm^{-2} ster^{-1} or 6.4×10^7 photons s^{-1} cm^{-2} ster^{-1}). The variation of the background light of the night sky, $B(\lambda)$, with λ is shown in figure 2.5.

2.4.2.5 Collection area

The unique feature of gamma-ray air shower detectors is that because the secondary radiation arrives at detector level as a broad but thin disk, a simple detector can have a large collection area for air shower detection, i.e. for primary gamma-ray detection. This is independent of the mirror collection area (which only determines the minimum gamma-ray energy that can be detected (see earlier)). Since the radius of the Cherenkov light pool on the ground is ~120 m, the shower detection area is ~5×10^4 m^2. It is the ability to achieve this large collection area (huge by astronomical detector standards), while retaining good angular and energy resolution, that makes the atmospheric Cherenkov technique so powerful. In almost all other wavebands, the collection area is determined by the dimensions of the telescope: such collection areas are still very small (although they can be enhanced by the use of focusing optics). Since the cosmic gamma-ray fluxes are low at high energies, this large collection area is essential for the discipline to be viable.

2.5 The background of cosmic radiation

2.5.1 Charged cosmic rays

So far there has been no discussion of the most troublesome background of all, cosmic rays, which are the limiting factor in gamma-ray observations. The charged cosmic radiation, over the energy range of interest to the VHE gamma-ray astronomer, is 10^3–10^4 times as numerous as the diffuse gamma-ray background. Cosmic ray ions, mostly protons, interact in the upper atmosphere and initiate a particle cascade which is superficially similar to an electromagnetic cascade. Helium ions, although only 10% of the cosmic rays, are an additional background component. The Cherenkov light distribution from these hadronic showers is similar to that from gamma-ray showers so that simple first-generation Cherenkov gamma-ray telescopes are unable to distinguish between the two. This would not be too serious if it were not for the fact that, in the field of view of a simple telescope whose solid angle is optimized for gamma-ray detection, the background of cosmic ray events is 10^3 times as numerous as the strongest steady gamma-ray discrete source thus far detected! Because of interstellar magnetic fields, the arrival directions of the charged cosmic rays are isotropic; hence a discrete source of gamma rays can stand out only as an anisotropy in an otherwise isotropic distribution of air showers. Unfortunately, a gamma-ray source would have to be very strong (a few per cent of the cosmic radiation) to be detectable in this way. Since the cosmic ray flux is pervasive, isotropic, and time invariant at these energies, it is impossible to avoid and it might seem impossible to do gamma-ray astronomy in this way. Fortunately, there are a number of factors concerning the properties of hadronic showers and pure electromagnetic cascades that make the ground-based study of cosmic sources of VHE gamma rays with Cherenkov telescopes possible.

There are very distinct differences in the cascade development in electromagnetic and hadronic cascades [14, 10]. The electromagnetic cascade, initiated by an electron or photon, has been described earlier. It consists almost entirely of electrons, positrons, and photons. The hadronic cascade is initiated by a charged ion and the core of the cascade consists of the products of hadronic interactions [6]. These feed lesser electromagnetic cascades whose products are largely responsible for the emission of Cherenkov light. Because a greater proportion of the energy in an electromagnetic cascade goes into particles that are efficient at initiating Cherenkov light, the typical Cherenkov light yield is two to three times that of a primary cosmic ray of the same energy.

The hadron interactions in the core emit their secondary products at wider angles of emission than their electromagnetic counterparts, so that the hadronic cascade is broader and more scattered. This can be seen in figure 2.1 where the Monte Carlo simulated development of a shower is depicted. The resulting Cherenkov light distribution in the focal plane of a detector is broader than that from a gamma-ray-initiated air shower and provides a simple method for differentiating between the two. Some of the secondary particles emitted from the core are penetrating particles which can reach ground level. These, as well as the larger fluctuations in the development of the hadron shower, have the effect of increasing the fluctuations in the Cherenkov shower image. Also, because they are local, the light that they radiate in the ultraviolet part of the spectrum of Cherenkov light is relatively unabsorbed.

The time spread of the Cherenkov light pulse from the hadronic shower is somewhat longer than that from the pure electromagnetic cascade, since the penetrating particles (and their local Cherenkov light) arrive early.

Other effects, such as polarization, are also different but it has not proved practical to use them as discriminants.

The cartoon in figure 2.7 illustrates the four most popular discriminants that have been used to differentiate gamma-ray showers from the hadronic background showers: lateral distribution, time spread, light spectrum, and angular distribution. The last of these has proven to be the most effective and is discussed in the next section.

The cosmic electron background is a factor of 100–1000 times less than the background due to hadronic cosmic rays. Since the electrons also produce electromagnetic cascades, they constitute a small, but virtually irreducible,

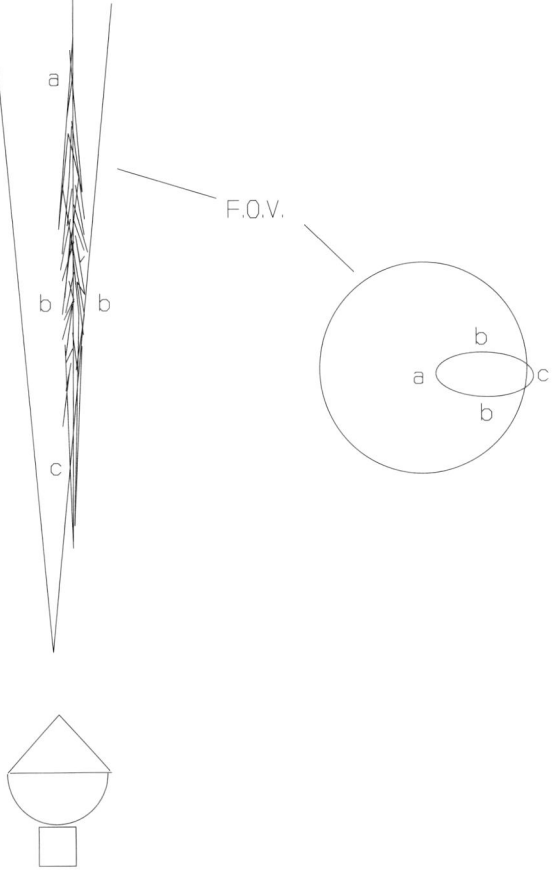

Figure 2.7. The geometry of the Cherenkov light images from an air shower (on left) as recorded by a camera on an atmospheric Cherenkov telescope (on right). The shower is parallel to the optic axis of the telescope and is inverted here. The light in the image comes from the top of the shower (a), from the middle (b) and from the bottom (c).

background. For the next generation of more sensitive ACTs, they may constitute the limiting factor in the energy range around 300 GeV.

2.5.2 Flux sensitivity

Ground-based gamma-ray telescopes operate in a domain where their flux sensitivity is dominated by an unavoidable background of cosmic ray events. The cosmic ray background has a power-law spectrum:

$$F_{cr}(> E) \propto E^{-a}.$$

In the range of interest, $a = 1.7$. Similarly, the gamma-ray source energy distribution can be assumed to have the form:

$$F_\gamma(> E_\gamma) \propto E_\gamma^{-a_\gamma}.$$

a_γ can have values from 1 to 3 and is generally assumed to increase with energy, i.e. the spectrum steepens.

If S equals the number of gamma rays detected from a given source in a time, t, and A_γ is the collection area for gamma-ray detection, then $S = F_\gamma(E)A_\gamma t$. The telescope will register a background, B, given by

$$B = F_{cr}A_{cr}(E)\Omega t$$

where $A_{cr}(E)$ is the collection area for the detection of cosmic rays of energy E.
Then the standard deviation,

$$\sigma \propto S/B^{1/2} \propto E^{1.7/2-a_\gamma}[A_\gamma/A_{cr}]^{1/2}t^{1/2}.$$

The minimum number of standard deviations, σ, for a reliable source detection is generally taken as five.

2.6 Atmospheric Cherenkov imaging detectors

2.6.1 Principle

The development of the Cherenkov imaging technique gave the first effective discrimination of gamma-ray showers from background hadron showers [26, 27]. An array of PMTs in the focal plane of a large optical reflector constitutes a camera and is used to record a Cherenkov light picture of each air shower. The camera is triggered when a preset number (usually two or three) of the PMTs detect a light level above a set threshold within a short integration time. The light level in all pixels is then recorded digitally and the image is analyzed offline to determine whether it has the expected characteristics of a gamma-ray shower with a point of origin at the center of the field of view. Discrimination against the background of charged cosmic ray showers is based on two factors:

(a) *geometry*—showers which arrive parallel to the optic axis (the putative direction of the source) will have roughly elliptical images which appear to radiate from the center of the camera;
(b) *physics*—as discussed in section 2.5.1, the image from a hadronic shower will be broader and more irregular than the image from an electromagnetic shower (figure 2.8).

It is fortunate that property (b) helps in the definition of (a).

The recorded optical image can be characterized using moment fitting by a few simple parameters, e.g. the width and length of the roughly elliptical images

Atmospheric Cherenkov imaging detectors

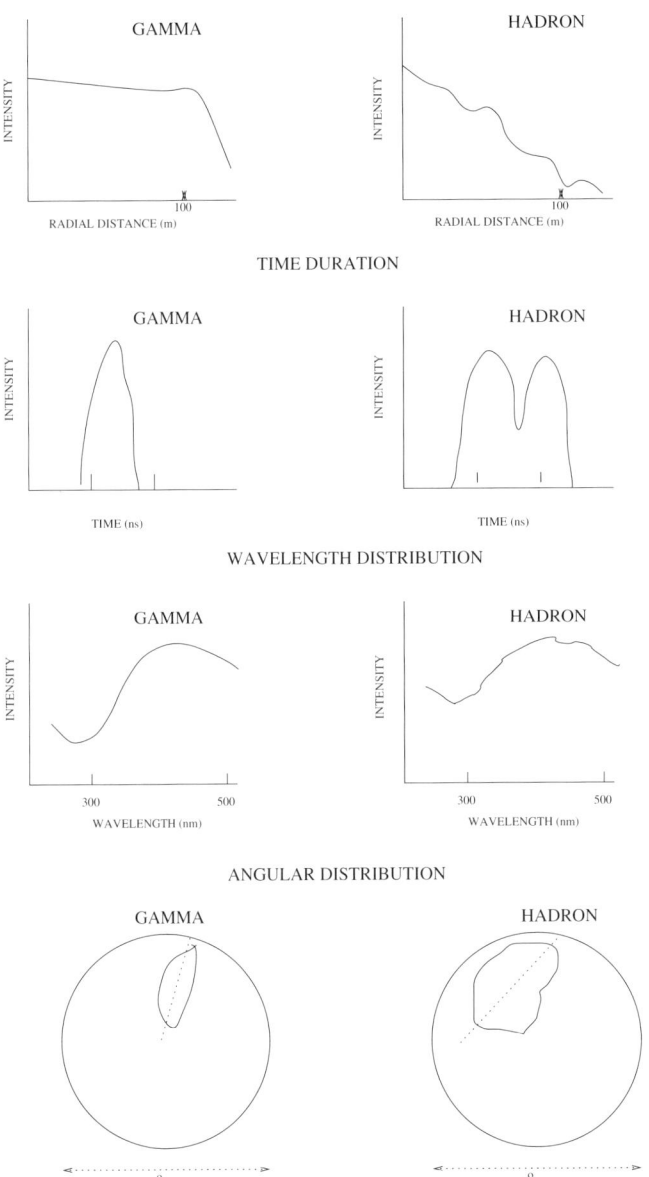

Figure 2.8. The different distributions of various parameters for gamma (left) and hadron (right) showers. (1) Radial distribution of light about the shower core at detector level. (2) Distribution of times of arrival of optical photons at detector. (3) Cherenkov light as a function of wavelength. (4) Distribution of light in angular space.

[9]. The orientation of the major axis of the ellipse will be different for gamma-ray images coming from a discrete source on the optic axis and that from randomly oriented images from the isotropically distributed background cosmic ray light images.

2.6.2 Angular resolution

The good angular resolution of atmospheric Cherenkov detectors arises from the inherent property of electromagnetic interactions that the emission angle of secondary particles is small. Thus the trajectory of the core of the shower is very close to the trajectory that the primary gamma ray would have followed if it had not interacted. At lower energies, Coulomb scattering of the secondary electrons becomes an additional consideration. The centroid of the light spot, from a shower parallel to the optic axis, which hits the ground 100 m from it, is displaced from the center of the field of view of the detector by $\sim 1°$. Hence first-generation ACTs had fields of view of 1–2° to get the maximum collection area. To improve the angular resolution, the trajectory of the shower is measured by recording the image of the shower (figure 2.9). This method is favored because with a single detector the arrival direction of the shower can be determined to 0.1°; with an array of detectors it can be fixed to 0.05°. The position of a source from which a few hundred gamma rays are detected can then be fixed to a few arc-min. It is remarkable that the source position can be pinpointed to this accuracy, given that the detector is some 20 km away from the transformation of the gamma ray into an electron–positron pair.

2.6.3 Energy resolution

The number of secondary particles in an air shower at shower maximum is proportional to the energy of the primary photon over a wide range of energies. Since the height of the shower maximum also varies with energy, this property is only useful as a measure of shower energy if the particle density can be sampled at various heights. However, most of these particles cause the emission of Cherenkov light which is beamed in the forward direction and which suffers little attenuation; the Cherenkov light received by a ground-based detector is, thus, a good measure of the total number of particles and hence of the primary energy. The chief uncertainty in the measurement is the distance to the shower core. If the measurement is made in the region from 50 to 130 m from the shower core, the effect of distance uncertainty is small. A single imaging detector can achieve an energy resolution of 30–40%, and an array of parallel detectors an energy resolution of 10–15%.

2.6.4 Existing imaging telescopes

There are currently seven observatories around the globe using variants of the atmospheric Cherenkov imaging (ACT) technique (table 2.2). A modern version

Table 2.2. Existing and planned ACT observatories.

Experiment	Location	Elevation (km)	Collectors	Mirror area (m^2)	Pixels/ camera	Energy (GeV)	Ref.
Whipple	Arizona, USA	2.3	1	75	467	250	[3]
GT-48	Crimea, Ukraine	0.6	2	27	37	1000	[24]
CANGAROO	Woomera, Australia	0.2	1	75	256	400	[7]
SHALON	Tien-Shan, Russia	3.3	1	10	144	1000	[18]
HEGRA	La Palma, Spain	2.2	6	9	271	500	[15]
CAT	Pyrenees, France	1.6	1	18	600	250	[20]
TACTIC	Mt Abu, India	1.3	4	10	349	500	[2]

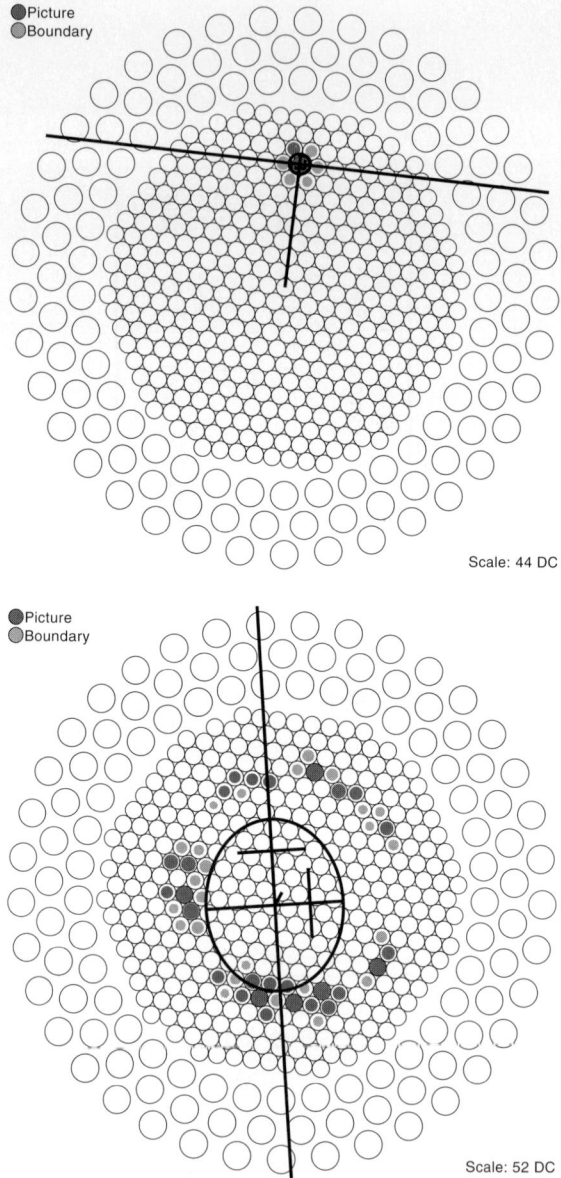

Figure 2.9. Some typical images recorded by a Cherenkov light camera: top, gamma-ray image; bottom, cosmic ray image; opposite top, sky noise trigger; opposite bottom, part of muon ring image. (Figure: S Fegan.)

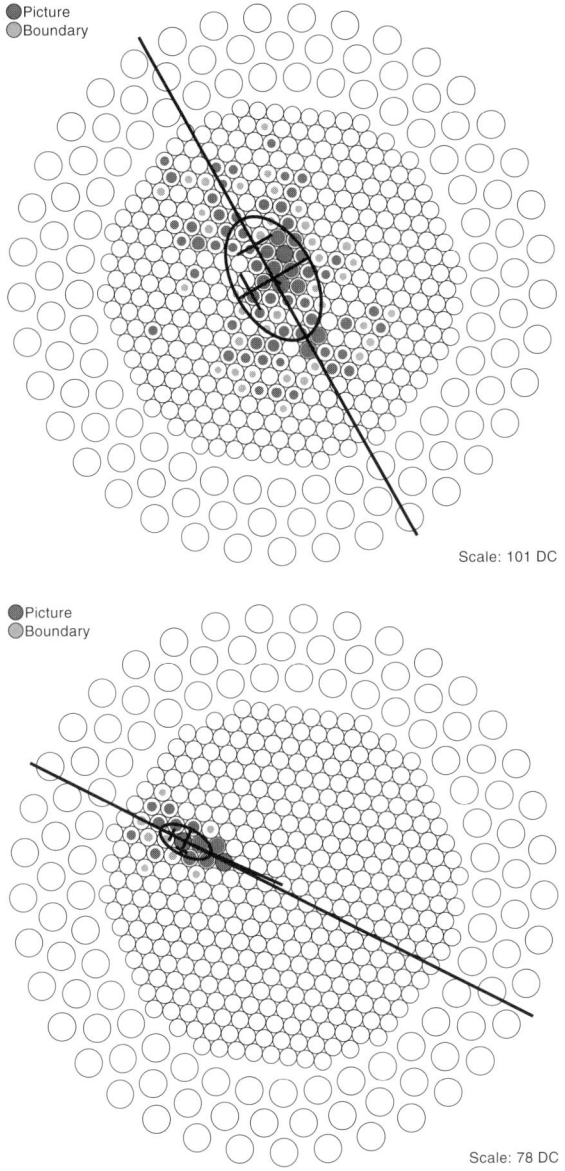

Figure 2.9. (Continued.)

of a camera (in use at the Whipple Observatory) is shown in figure 2.10. Background rejection of cosmic rays is in excess of 99.7%, and the technique is effective from energies of 250 GeV to 50 TeV. A signal with significance of

Figure 2.10. The 490 pixel camera of the Whipple Observatory made up of individual photomultipliers. The 379 inner pixels each subtend an angle of $0.12°$; the 111 outer pixels each subtend an angle of $0.25°$.

5–10σ can be detected from the Crab Nebula in just an hour of observation. The largest single imaging telescope currently under construction is the 17 m European telescope, called MAGIC [16], which will be located on La Palma in the Canary Islands, Spain. It is hoped to reach an energy threshold as low as 30 GeV with this telescope because of the large aperture and new technologies that will be incorporated into this ambitious instrument.

2.6.5 Arrays

The atmospheric Cherenkov imaging technique can be significantly improved by the use of multiple telescopes with separations of the same order as the lateral spread of the light from the shower. Early on in the development of the Cherenkov technique, it was recognized that the stereo detection of the shower would improve the angular resolution and permit the rejection of background cosmic ray air showers. Multiple images of the same shower offer many advantages, such as reduced energy threshold by using a coincident trigger between telescopes, improved hadron discrimination from multiple image characterization, elimination of local muon background, shower axis

Figure 2.11. The HESS array of four 12 m telescopes in Namibia. The first telescope came into operation in 2002.

location, determination of shower maximum, and better angular resolution. This approach was first demonstrated by the Armenian–German–Spanish collaboration, HEGRA, with five small telescopes on La Palma in 1997 [15].

The exciting advances made by the present generation of imaging telescopes justify the construction of arrays of large imaging telescopes. Such systems must have the following properties:

- *Large effective area*: >0.1 km^2 to provide sensitive measurements of short variability time scales.
- *Better flux sensitivity*: detection of sources which emit gamma rays at levels of 0.5% of the flux from the Crab Nebula at energies of 200 GeV in 50 hr of observation.
- *Reduced energy threshold*: an effective energy threshold of <100 GeV with significant sensitivity at 50 GeV.
- *Improved energy resolution*: an RMS spectral resolution of $\Delta E/E < 0.15$ over a broad energy range.
- *Increased angular resolution*: <0.05° for individual photons; source location capability better than 0.005°.
- *Large field of view (FOV)*: at least 3° diameter as used in many current ACTs.

The next generation of ACTs will see the construction of three arrays of large telescopes (table 2.3): an Irish–UK–USA collaboration that is building an array of seven telescopes in Arizona (VERITAS); an Australian–Japanese collaboration that is building four telescopes in Australia (CANGAROO-III); and a largely European collaboration that is building an array of initially four, and eventually 16, telescopes in Namibia (HESS) (figure 2.11).

Table 2.3. Next-generation ACT arrays.

Experiment	Location	Elevation (km)	Collectors	Mirror area (m²)	Pixels/ camera	Energy (GeV)	Ref.
HESS	Gamsberg, Namibia	1.8	4 (16)	100	960	50	[12]
VERITAS	Arizona, USA	1.5	7	100	499	50	[28]
CANGAROO-III	Woomera, Australia	0.2	4	75	577	50	[17]

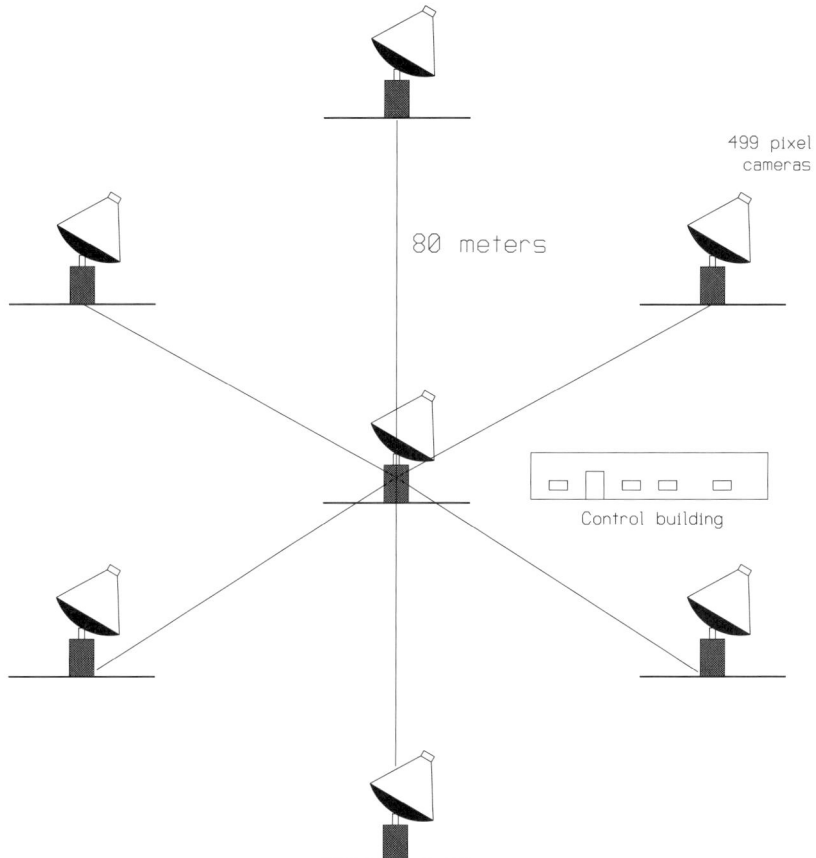

Figure 2.12. The layout of the seven 12 m telescopes that will comprise the VERITAS observatory in southern Arizona.

The Very Energetic Radiation Imaging Telescope Array System (VERITAS) was the first of these next-generation telescopes to be proposed. The seven telescopes in VERITAS will be identical and will have the geometrical layout shown in figure 2.12. Six telescopes will be located at the corners of a hexagon of side 80 m and one will be located at the center. The telescopes will each have a camera consisting of 499 pixels with a field of view of 3.5° diameter. The flux sensitivity of VERITAS (which will be very similar to that of HESS and CANGAROO-III) is given in the next chapter (figure 3.7) where it is contrasted with that of existing telescopes, both ground-based and space-borne.

More by accident than design, the next generation of major new telescopes will have a logical distribution in latitude and longitude with MAGIC and VERITAS in the Northern Hemisphere and HESS and CANGAROO-III in the

Southern Hemisphere. Unfortunately, the other telescopes (including the air shower arrays, discussed below) are all concentrated in the Northern Hemisphere.

2.7 Other ground-based detectors

2.7.1 Particle air shower arrays

At high energies (>10 TeV), there are sufficient particles reaching ground level that the shower can be detected, its energy estimated and its arrival direction determined. This requires large arrays of particle detectors through which some of the particles must pass. Typically these experiments have angular resolutions of 1°, energy resolutions of 30%, and collection areas in excess of 10 000 m^2. It is difficult to discriminate gamma-ray showers from hadron showers in this way and hence this field has been slow to develop. Despite this limitation heroic efforts were made in the 1980s to develop this form of gamma-ray astronomy and considerable resources were devoted to this endeavor. These experiments had energy sensitivity in the range 10^{14}–10^{16} eV; they attempted to use the ratio of penetrating particles (muons) to electrons as the gamma-ray discriminant. The largest of these were the Cygnus array at Los Alamos, New Mexico, and the CASA Array at Dugway, Utah. Although there were a number of possible detections (e.g. Cygnus X-3 [22]), they were never verified and interest in this part of the energy spectrum appears to have waned.

There are now efforts to reduce the energy threshold of these experiments to overlap with those of imaging ACTs so that they can operate in an energy band where there are known sources. Two air-shower particle detectors have successfully detected gamma rays of a few TeV from the strongest sources. One is a large water Cherenkov detector, Milagro, near Los Alamos, New Mexico, USA, at an elevation of 2.6 km [23]. The other is a densely packed array of scintillation detectors in Tibet, which operates at an elevation of 4.3 km [1]. Although these telescopes are somewhat less sensitive, they have the advantage over Cherenkov telescopes that they can operate continuously and hence monitor a large section of the sky.

2.7.2 Solar power stations as ACTs

An alternative approach to the detection of gamma rays using the Cherenkov light emission in the atmosphere is the use of the large arrays of optical heliostats built for solar energy power stations as light collectors. Generally, these arrays are no longer in use, although, in principle, they can be used at night without interfering with the solar energy generation activity. Although the optics and location of these facilities are not ideal, they do offer very large mirror collection areas and, hence, the possibility of low energy thresholds (30–100 GeV). At these energies the hadronic showers are much less efficient at producing Cherenkov light and, hence, this troublesome background will virtually disappear. There is the possibility of

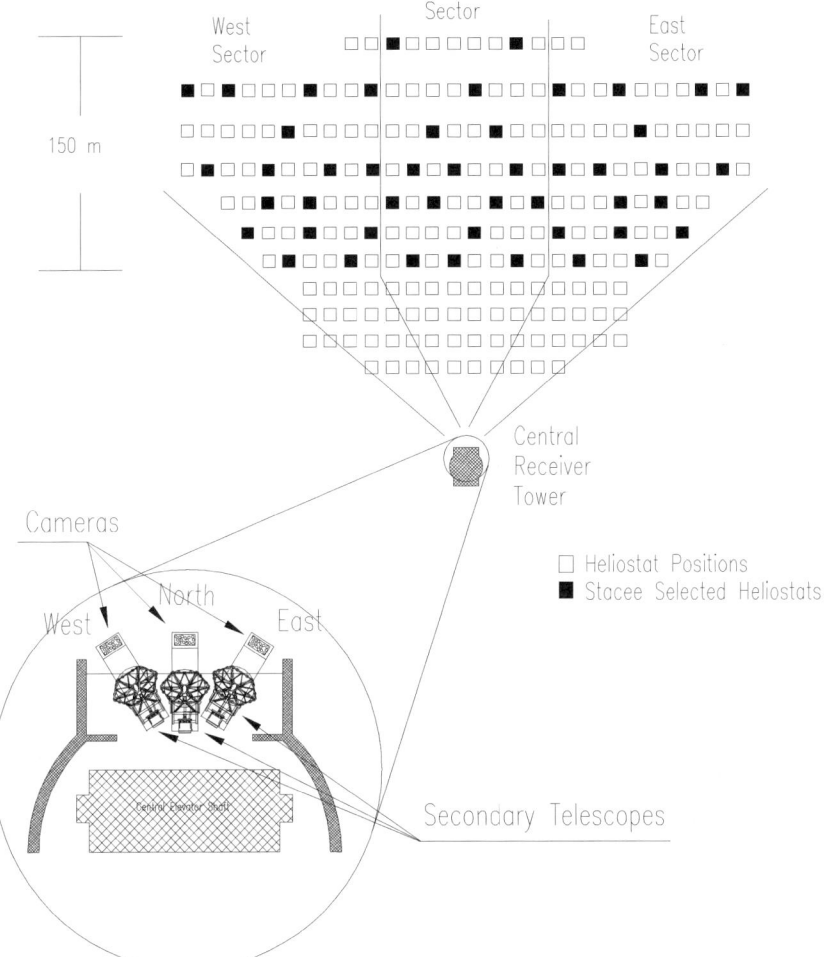

Figure 2.13. A schematic diagram of the Albuquerque solar collector as used for gamma-ray astronomy in the STACEE experiment. (Figure: R Ong.)

measuring the lateral distribution of the light and using that as a discriminant. The field of view of these detectors is of necessity small so that these systems are not suitable for observing extended sources. They may, however, make a significant contribution to the study of sources with steep spectra, e.g. pulsars, distant AGN, etc.

These techniques are not easy. The focal plane instrumentation is complex since each heliostat or group of heliostats must be focused onto individual photomultipliers. There are currently three experiments in operation: STACEE (Albuquerque, New Mexico, USA) [5], Solar Two (Barstow, California, USA)

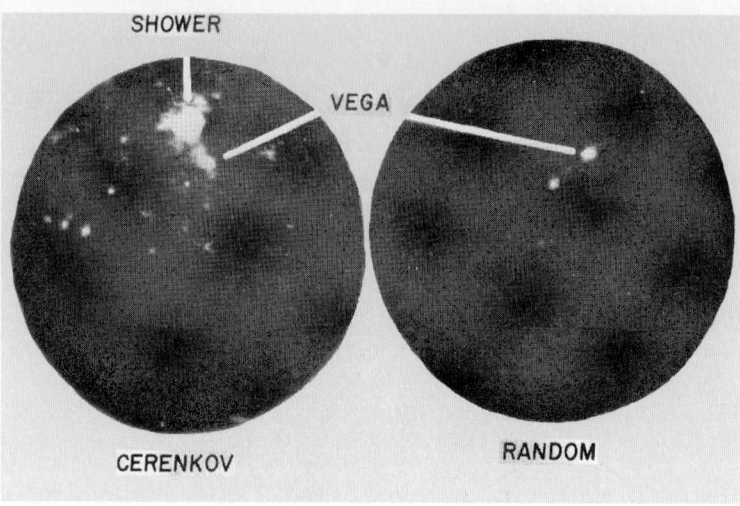

Figure 2.14. The first image intensifier picture of Cherenkov light from a large cosmic ray air shower [14]. (Photo: N A Porter.)

[29], and CELESTE (Pyrenees, France) [21]. The STACEE detector uses 48 heliostats, each of area 37 m^2. The secondary optics requires three spherical mirrors of aperture 2 m (figure 2.13). Some of the sources seen with imaging detectors at energies above 300 GeV have been seen with these solar detectors at energies below 100 GeV.

Historical note: Cherenkov images

The first direct images of Cherenkov light from air showers were recorded in 1960 by Hill and Porter [8] using an image intensifier camera system coupled to a small mirror. The image intensifier system was adapted from a particle experiment and involved three stages of image intensification. It was electronically gated on by the arrival of a Cherenkov light pulse at a parallel light detector (a 12.5 cm PMT). The camera was mounted at the focus of a 30 cm mirror; it had an effective time exposure of 10 μs. Cherenkov light images were recorded on photographic film with angular sizes of a few degrees. The threshold energy of the cosmic rays was estimated as >500 TeV. In one of the images (figure 2.14) the star, Vega, is visible giving some idea of the brightness of the images. Although these images are of poor quality, they indicated the potential information that was available in Cherenkov light shower images, i.e. arrival direction, nature of primary, energy. Image intensifier systems did not prove practical for VHE gamma-ray astronomy studies because of their slow readout and limited size, which made coupling to large optical systems inefficient; these were to be replaced by cameras with arrays of PMTs.

References

[1] Amenomori M *et al* 1997 *Proc. 25th ICRC (Durban)* vol 5, ed M S Potgeier *et al* (Potchefstroom University) p 245
[2] Bhat C L *et al* 1997 *Proc. Workshop on VHE Gamma Ray Astronomy (Kruger Park) (August)* (Potchefstroom University) p 196
[3] Cawley M F *et al* 1990 *Exp. Astron.* **1** 173
[4] Cawley M F and Weekes T C 1995 *Exp. Astron.* **6** 7
[5] Chantell M C *et al* 1998 *Nucl. Instrum. Methods* A **408** 468
[6] Galbraith W 1958 *Extensive Air Showers* (London: Butterworth Scientific)
[7] Hara T *et al* 1993 *Nucl. Instrum. Methods* A **332** 300
[8] Hill D A and Porter N A 1960 *Nature* **191** 690
[9] Hillas A M 1985 *Proc. 19th ICRC (La Jolla)* **3** 445
[10] Hillas A M 1996 *Proc. 'TeV Gamma-ray Astrophysics' (Heidelberg) (Space Sci. Rev. 75)* ed H J Volk and F A Aharonian (Dordrecht: Kluwer) p 17
[11] Hillas A M 2002 Private communication
[12] Hofmann W 1999 *GeV–TeV Astrophysics: Towards a Major Atmospheric Cherenkov Detector IV (Snowbird, Utah)* ed B L Dingus, M H Salamon and D B Kieda (New York: AIP) p 500
[13] Jelley J V 1958 *Cherenkov Radiation* (New York: Pergamon Press)
[14] Jelley J V and Porter N A 1963 *Quart. J. R. Astron. Soc.* **4** 275
[15] Konopelko A *et al* 1999 *Astropart. Phys.* **10** 275
[16] Lorenz E 1999 *GeV–TeV Astrophysics (Snowbird, Utah) (AIP Conf. Proc. 515)* ed B L Dingus, M H Salamon and D B Kieda (New York: AIP) p 510
[17] Matsubara Y 1997 *Towards a Major Atmospheric Cherenkov Detector (Kruger Park)* ed O C de Jager (Potchefstroom University) p 447
[18] Nikolsky S I and Sinitsyna V G 1989 *Proc. Workshop on VHE Gamma Ray Astronomy, Crimea (April)* ed A A Stepanian, D J Fegan and M F Cawley (Crimean Astrophysical Observatory) p 11
[19] Ong R A 1998 *Phys. Rep.* **305** 93
[20] Punch M 1995 CAT collaboration *Towards a Major Atmospheric Cherenkov Detector IV (Padova)* ed M Cresti (University of Padova) p 356
[21] Quebert J *et al* 1995 *Towards a Major Atmospheric Cherenkov Detector IV (Padova)* ed M Cresti (University of Padova) p 248
[22] Stamm M and Samorskii W 1983 *Astrophys. J. Lett.* **268** L17
[23] Sinnis G *et al* 1995 *Nucl. Phys. B (Proc. Suppl.)* **43** 141
[24] Vladimirsky B M *et al* 1989 *Proc. Workshop on VHE Gamma Ray Astronomy, Crimea (April)* ed A A Stepanian, D J Fegan and M F Cawley (Crimean Astrophysical Observatory) p 21
[25] Weekes T C 1988 *Phys. Rep.* **160**
[26] Weekes T C and Turver K E 1977 *Proc. 12th ESLAB Symp. (Frascati)* ed R D Wills and B Battrick (European Space Agency) ESA SP124, p 279
[27] Weekes T C *et al* 1989 *Astrophys. J.* **342** 370
[28] Weekes T C *et al* 2002 *Astropart. Phys.* **17** 221
[29] Zweerink J A *et al* 1999 *26th ICRC (Salt Lake City)* **5** 223

Chapter 3

High energy gamma-ray telescopes in space

3.1 Introduction

It is not practical to consider the detection of gamma rays of 10 GeV or less using ground-based detectors. In fact, there have been few observations as yet in the energy interval 10–100 GeV. To go to lower energies, it is necessary to carry the telescope above the atmosphere. Gamma rays of these energies cannot be reflected so the collection area is only as large as the detector. In practice, it is much less than this since the troublesome background of charged cosmic rays must be screened out. Since the gamma-ray fluxes are small, long exposures are necessary. Early experiments in the HE band used balloons, unlike x-ray astronomy where the pioneering experiments were carried out from rockets. Since 1973 almost all measurements have come from telescopes carried on satellites. Unfortunately, space missions are expensive and flight opportunities are few; there is not an international agreement to dovetail missions so that there are significant interruptions in coverage, e.g. there were no HE telescopes at all in orbit from 1985 to 1991 and from 2000 to, at least, 2003.

The detection techniques employed in space telescopes are determined by the dominant interaction process in the energy region of interest. The most important process in the HE region is pair production but for completeness we also consider briefly the Compton region which is most appropriate for ME gamma-ray astronomy.

3.2 Pair production telescopes: high energy

The early balloon-borne experiments generally used spark chambers as their principal detector element [4]. In fact, the spark chamber, long obsolete for high energy physics experiments, has been the workhorse detector for HE gamma-ray astronomy in the energy range 30 MeV to 10 GeV from the early 1960s through to the end of the century (figure 3.1). The three gamma-ray satellite experiments, which provided almost all the results during this period, used the spark chamber

Pair production telescopes: high energy

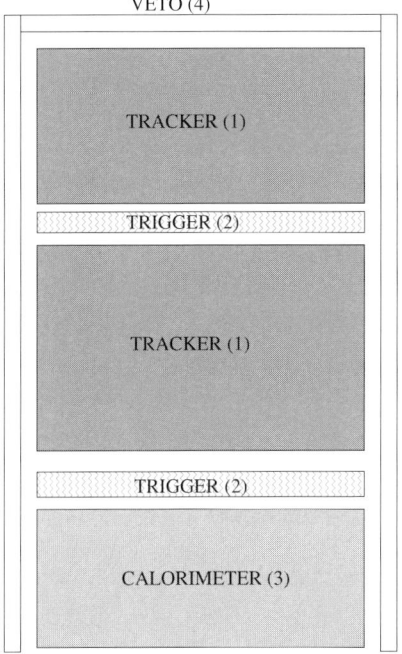

Figure 3.1. The four basic elements of the spark chamber pair production telescope: (1) the tracking spark chamber; (2) the trigger; (3) the calorimeter; (4) the anti-coincidence detector.

as their principal detector. These were the USA's SAS-2 (1973), Europe's COS-B (1975–82) and the joint European–USA EGRET on the Compton Gamma Ray Observatory (CGRO) (1991–2000).

Although the basic principles of the pair-production telescope are simple, the detailed design is complex and accounts for the fact that the effective collection area is often far smaller than the geometrical cross section of the telescope. This is illustrated by EGRET, the pair production telescope on the CGRO. As with most pair production spark chamber telescopes, this consisted of four distinct components which are discussed here and shown schematically in figure 3.2 [8].

(1) The Tracker: the spark chamber usually consists of a series of parallel metal plates in a closed container. The alternate plates are connected together electrically with one set permanently connected to ground. Upon an indication that a charged particle has passed through the chamber, a high voltage is applied to the second set of plates. The chamber contains a gas at a pressure such that the ionization left behind by the passage of the charged particle causes an electric spark discharge between the plates. The gas is a mixture of neon and ethane. An electron pair created by a gamma-ray interaction in one of the plates

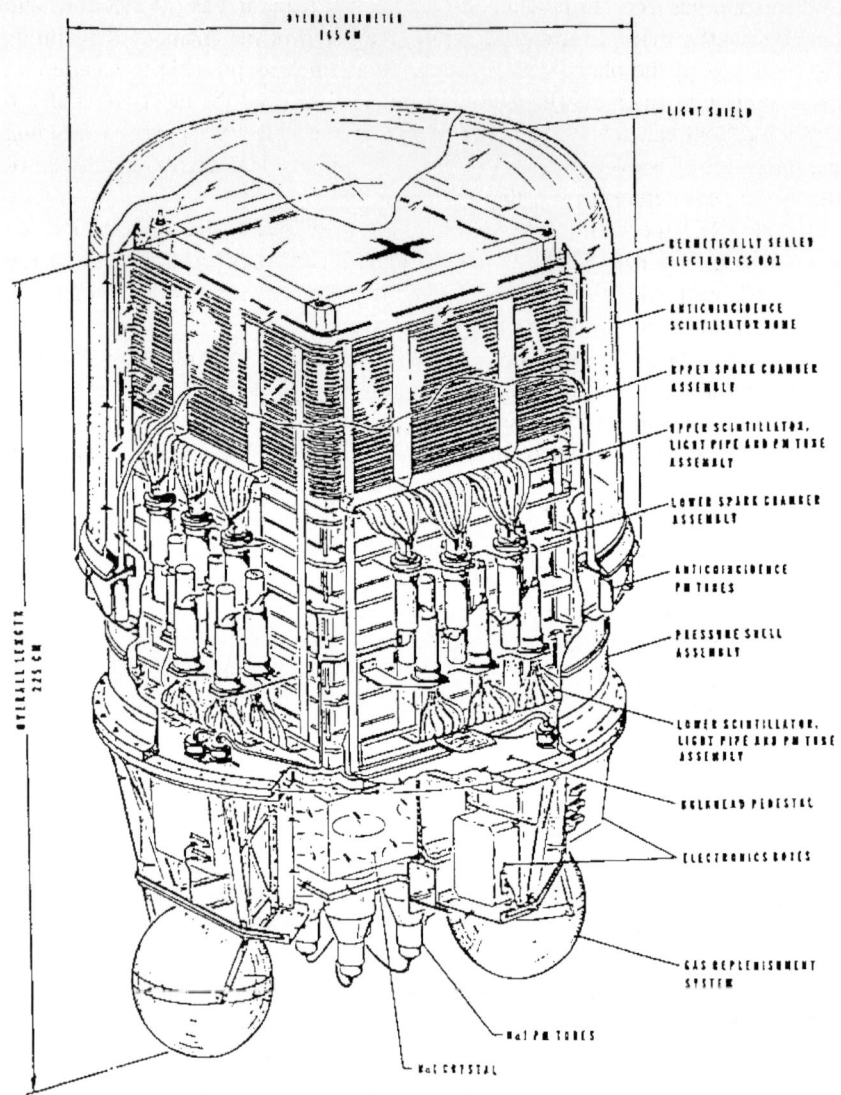

Figure IV-1. The EGRET Instrument

Figure 3.2. Example of a spark chamber telescope: EGRET on the Compton Gamma Ray Observatory. The telescope was sensitive from 30 MeV to 30 GeV. The field of view was ±20° and the energy resolution was about 20%. It operated from 1991 to 2000 by which time it had exhausted its gas supply. (Figure: D Thompson).

is then readily apparent as a pair of sets of sparks that delineate the path of the electron and positron. In practice, the tracks are disjointed as the electrons and positrons suffer multiple scattering within the plates of the chamber. This limits the thickness of the plates which should be as thick as possible to ensure that the gamma rays interact effectively but not so thick that the electrons undergo excessive Coulomb scattering in the plate material. Multiple plates ensure that the tracks are effectively mapped. The collection area and angular resolution of the telescope is determined by the spark chamber geometry.

In EGRET the spark chamber 'plate' consisted of 28 wire grids interleaved with interaction plates of 0.02 radiation length thickness in which the gamma ray interacted. Each wire was threaded through a magnetic core memory, which was read out and reset after each event.

(2) The Trigger: At least one electron must emerge from the spark chamber to ensure that it causes a trigger that initiates the application of the high voltage pulse to the second set of plates to activate the spark chamber. A permanent high-voltage difference cannot be maintained between the plates, as the spark discharges would then take place spontaneously.

EGRET was triggered by a coincidence between two thin sheets of plastic scintillator with a 60 cm separation (sufficient to recognize and reject upward-going charged particles). It was the need for this trigger which limited the lower energy threshold of the spark chamber telescope. The trigger detection system effectively defines the field of view of the telescope.

(3) The Calorimeter: The electrons must be completely absorbed if their energy is to be measured; to achieve this there must be a calorimeter that is some radiation lengths thick.

In EGRET, as in most spark chamber telescopes, this was a NaI(Tl) crystal, whose sole function was to measure the total energy deposited. At the low end of the sensitivity range, the energy of the electrons could also be determined by the amount of Coulomb scattering in the plates of the spark chamber. The calorimeter had no directional properties but could act as an independent gamma-ray burst detector (see chapter 13).

(4) The Veto: Finally the entire assembly is surrounded by an anti-coincidence detector which signals the arrival of a charged particle but which has a small interaction cross section for gamma rays. This consists of a very thin outer shell of plastic scintillator viewed by photomultipliers.

EGRET was the largest, and most sensitive, high energy gamma-ray telescope flown to date; it was the flagship instrument on the CGRO. Approximately the size of a compact car and with a total weight of 1900 kg, the telescope had an effective collection area of 1600 cm^2. Hence, despite its large weight and volume the collection area was not much larger than two of these pages. The characteristics of the telescope are listed in table 3.1.

The telescope was designed for a five-year lifetime. The gas which filled the chamber gradually became poisoned and had to be replenished. It was anticipated that a filling would last one year. Hence, only four gas canisters were attached to the instrument for replenishment at yearly intervals. In practice, the unprecedented and unexpected success of CGRO meant that the mission was extended as were the replenishment intervals so that for a considerable fraction of the nine-year lifetime of the mission, EGRET operated at less than optimum efficiency.

3.3 Compton telescopes

The usual detector of gamma rays in the difficult 100 keV to 10 MeV energy range is the scintillation detector, which consists of a solid or liquid material, in which light is produced by charged secondary particles resulting from the photoelectric or Compton scattering gamma-ray interaction (see appendix) and a photomultiplier tube (PMT), in which light is converted into an electrical signal. A common scintillation material is thallium-activated sodium iodide, NaI(Tl). Charged particles are rejected by surrounding the detector by another plastic scintillator detector. If the outer detector is shaped like a well with a small opening, then it can serve as a collimator with crude angular resolution. Many of the early detectors worked on this principle.

A more sophisticated detector is the Compton telescope in which two detectors are operated in series (figure 3.3). In the top scintillation detector, a primary gamma ray, which Compton-scatters in the forward downward direction, is selected (based on the energy registered by this first detector from the recoil electron); the gamma ray is then absorbed in another Compton scatter in the lower detector. The lower detector is surrounded by an anti-coincidence scintillator to veto charged particles coming up from below. Of necessity, because of the wide range of angles that the scattering may have, the efficiency of these simple detectors is poor, typically less than 1%. However, the energy and angular resolution is improved over the simple one-stage detector.

The upper detector should have a large cross section for Compton scattering over the desired energy range. In the 1–10 MeV region, the best material is one with low Z; hence, the detector should be a relatively thin liquid or plastic scintillator in which a single Compton scattering occurs with good efficiency. The second, lower, detector should totally absorb the product of the second Compton scatter and, hence, should be thicker and composed of high-Z material.

A double Compton scattering is also the basic principle used in the most sophisticated 1–30 MeV telescope flown to date, COMPTEL on the CGRO (figure 3.4). The primary gamma-ray incident within ±40° of the telescope axis was first Compton-scattered in the upper detector which was a low-Z liquid scintillator; the second scattering took place in the lower detector which was a high-Z NaI(Tl) scintillator. Each detector actually consisted of seven modules and the separation between the two layers was 1.5 m. Hence, time-of-flight was used to discriminate against upward-going particles. In addition, all of the detectors were surrounded by thin plastic anti-coincidence scintillators which

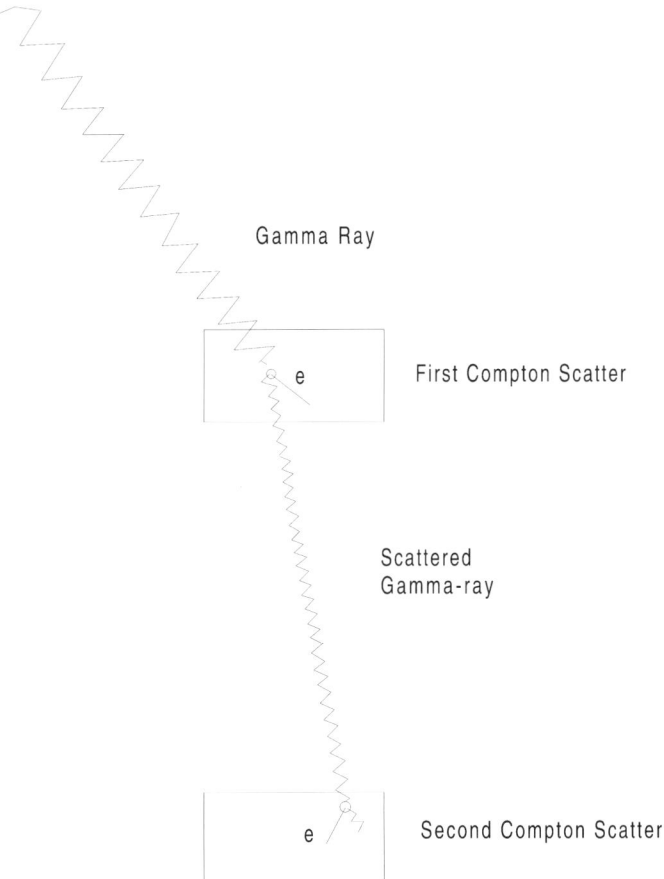

Figure 3.3. Schematic diagram of double Compton scatter. In the upper scintillator the primary gamma-ray Compton-scatters in the downward direction; the gamma ray is then absorbed in another Compton scatter in the lower detector. In each case the energy of the recoil electron is measured and, thence, the energy and arrival direction of the primary is determined.

responded to, and vetoed, charged particles. If the energy deposited in the upper and lower modules was measured, then the direction of the incident gamma ray was determined to be within a narrow ring on the sky and its energy estimated to about 5–10%. A source was then apparent as the locus of intersections of a number of such rings.

To constrain the incident gamma ray's energy and direction uniquely, the energies and directions of the recoil electron and scattered photon must be determined. To progress from source circles to small source error boxes,

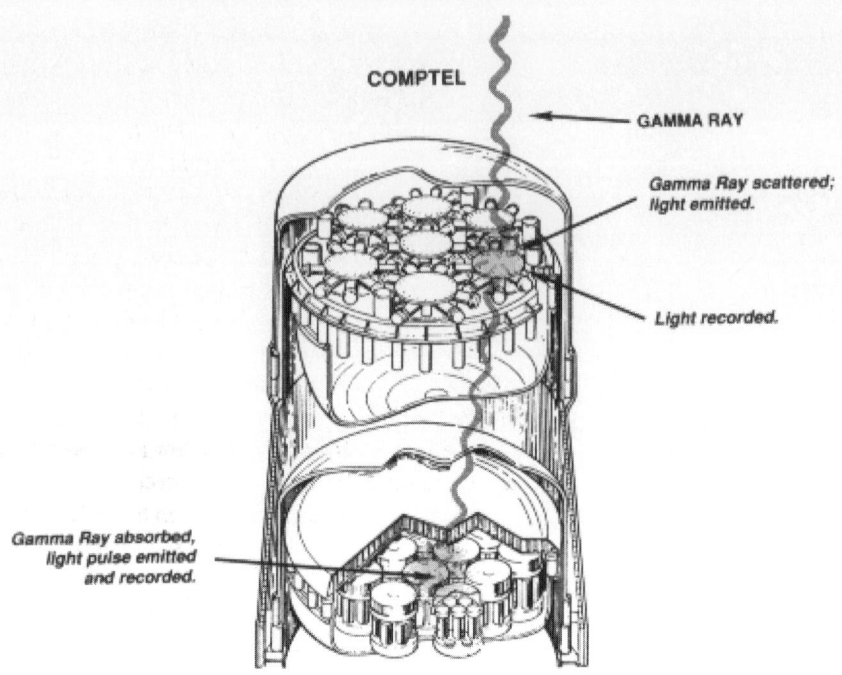

Figure 3.4. An example of a Compton telescope: COMPTEL. It was sensitive from 1 to 30 MeV. The angular resolution was 3–5°. (Figure: A Falcone.)

future Compton telescopes must have improved spatial and energy resolutions in both sets of detectors. In recent years, there have been considerable advances in detector technology, in particular in the development of position-sensitive semiconductor detectors. These include silicon strip detectors, room temperature cadmium zinc telluride detectors, and cooled germanium detectors. These telescopes will have angular resolutions of a few arc-min and energy resolutions of a few keV, a considerable improvement over COMPTEL.

3.4 Future space telescopes

3.4.1 INTEGRAL

The International Gamma-Ray Astrophysics Laboratory (INTEGRAL) is primarily a European mission [6]. It was selected by the European Space Agency (ESA) in 1993 as a medium size mission but has been modified somewhat from the original proposal because of the non-participation of the USA and UK in the mission. The telescopes operate in the 15 keV–10 MeV region with two prime objectives: good angular resolution (12 arc-min) and good energy resolution ($E/\delta E = 500$). To achieve these objectives two instruments are used: the Spectrometer SPI is based on solid-state germanium detectors with coded aperture masks to define the field of view; and the Imaging IBIS which uses cadmium telluride and caesium iodide detectors. These two gamma-ray telescopes are supplemented by small x-ray and optical telescopes for the monitoring of transient sources over a broad range of wavelengths. The detector design is based on the assumption that broad line emission is associated with point-like sources and narrow line emission comes primarily from extended sources as indicated by earlier missions. The mission was put into orbit on a Russian Proton rocket, on 17 October 2002. It is planned to have a two-year lifetime with possible extension to five years.

3.4.2 Swift

Set for launch in late 2003, Swift is a NASA MIDEX mission which is designed to give multi-wavelength coverage of gamma-ray bursts (see chapter 13) [3]. It will consist of three instruments: the Burst Alert Telescope, the Hard X-ray Telescope, and the UV and Optical Telescope. In principle, it will be capable of locating bursts to a few arc-sec. The Burst Alert Telescope will have five times the sensitivity of Burst and Transient Source Experiment (BATSE), the gamma-ray burst detector on CGRO, but will have an upper energy sensitivity limit of 150 keV. Hence, although billed as a gamma-ray mission, this is really an x-ray mission that will detect the soft end of gamma-ray bursts. Using the UV and Optical Telescope it will be possible to make onboard determinations of the redshift of the gamma-ray burst emitting galaxy.

3.4.3 Light imaging detector for gamma-ray astronomy (AGILE)

AGILE (light imaging detector for gamma-ray astronomy) is an Italian mission which will be launched in 2003 and which will carry a HE telescope, the Gamma-Ray Imaging Detector(GRID), as well as a hard x-ray telescope (Super-AGILE) [2]. GRID will cover the energy band from 30 MeV to 50 GeV with good spatial resolution and with a very wide field of view (3 sr). It will be the first HE gamma ray telescope to make use of silicon strip technology and, as such, it acts as a stalking horse for GLAST (see section 3.4.5). The tracker will have very good angular resolution. The calorimeter is only 1.5 radiation lengths thick and, hence, this instrument will have very limited energy resolution. As usual, the tracker and calorimeter will be surrounded by an anti-coincidence shield. Super-AGILE will cover the range 10–40 keV with 1–3 arc-min resolution; it consists of a thin layer of silicon strips with coded mask apertures mounted on the top of AGILE.

The remarkable features of this mission are that it is entirely an Italian effort, that it has been developed in a remarkably short time, that it will combine hard x-ray and gamma-ray detectors, and that the total weight is only 80 kg. Its

Table 3.1. Comparison of EGRET and GLAST.

Parameter	Units	EGRET (achieved)	GLAST (desired)
Energy range	MeV	20–30 000	20–300 000
Effective area	cm^2	1500	>8000
Field of view	sr	0.5	>2
Angular resolution	(100 MeV) degrees	5.8	3.5
	(>10 GeV) degrees		<0.15
Energy resolution	%	10	10
Source sensitivity	(>100 MeV) cm^{-2} s^{-1}	10^{-7} 1	0.06

sensitivity will be similar to that of EGRET so it cannot really be classified as a next-generation HE telescope.

3.4.4 Alpha Magnetic Spectrometer (AMS)

The Alpha Magnetic Spectrometer (AMS) is a very large and expensive mission designed to go on the International Space Station [1]. Its primary purpose is to study anti-matter but it also has some sensitivity for the study of cosmic ray isotopes and HE gamma rays. It will consist of a magnetic spectrometer with large acceptance angle, a large permanent magnet, layers of silicon tracker, scintillators, transition radiation detectors, and a solid state Cherenkov detector. For gamma-ray studies, it is hoped to have a sensitivity similar to that of EGRET. However, on the Space Station it will not be possible to direct the telescope so it will not be capable of responding to targets of opportunity. A preliminary version of the AMS was flown on the Space Shuttle in June, 1998. The AMS is scheduled to go on the Space Station in 2005.

3.4.5 The Gamma-ray Large-Area Space Telescope (GLAST)

The Gamma-ray Large-Area Space Telescope (GLAST) is the next-generation pair-production telescope; like AGILE it will replace the spark chamber with solid-state detectors which will be more compact, more efficient, and have better angular and energy resolution. However, the general principle of the telescope will be the same as EGRET, with an anti-coincidence scintillator shield, an interaction region with three-dimensional imaging system, and a calorimeter; the usual triggering system is unnecessary as the imaging system can be active all the time. There are no expendables, no noisy pulsed high voltage, and no sparks.

As in the past, the technology required for the new generation pair production telescope was developed largely for particle physics accelerator experiments and

Future space telescopes 51

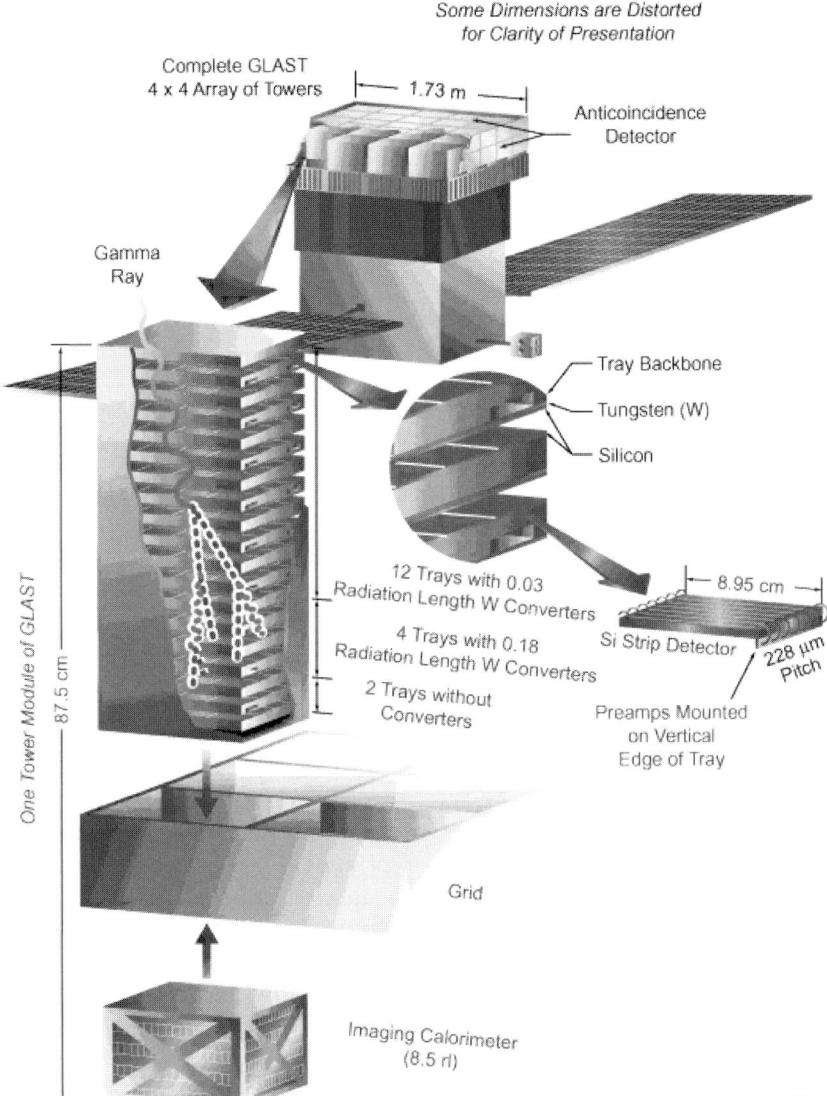

Figure 3.5. A schematic diagram of the Large Area Telescope on GLAST showing the three principal components: the tracking modules, the anti-coincidence detector, and the calorimeter. (http://glast.gsfc.nasa.gov/resources/)

was adapted for use in space. GLAST uses the silicon strip technology that has been used in high energy particle accelerator experiments for a number of years. It has not so far been used in space science applications but will be demonstrated in AGILE.

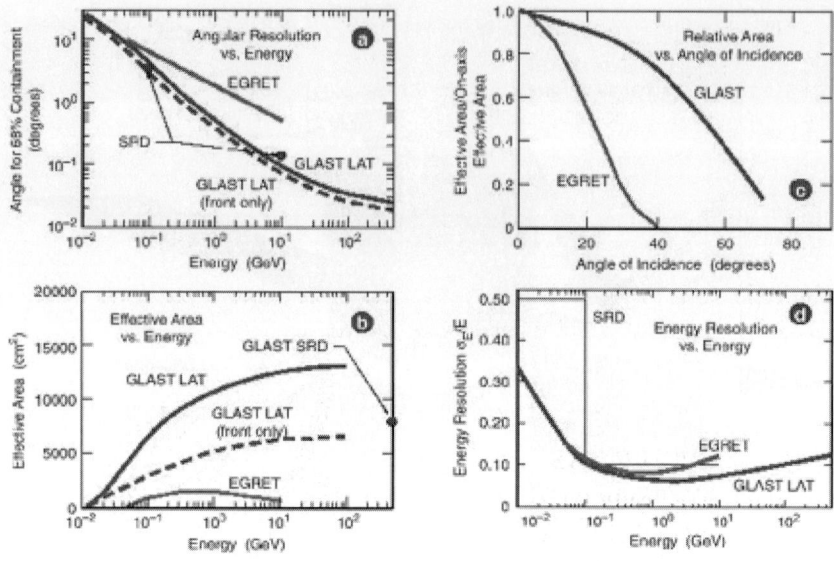

Figure 3.6. GLAST performance predictions as a function of energy compared with EGRET. (http://glast.gsfc.nasa.gov/resources/)

GLAST consists of two parts: the Large Area Telescope (LAT) and the Gamma-Ray Burst Monitor (GRM) [5]. GRM is a simple wide-field low energy instrument to alert GLAST to the occurrence of a gamma-ray burst. LAT has three components (figure 3.5).

(i) The tracker/convertor consists of 18 layers of ionizing particle-sensitive detectors with high Z. They provide the familiar pair production track which is used to distinguish the gamma rays from charged cosmic rays.

(ii) The calorimeter will be made of eight layers of bars of caesium iodide, with individual read-outs to give spatial resolution. The calorimeter will be 8.5 radiation lengths thick which permits the detector to operate with some efficiency up to 300 GeV.

(iii) The anti-coincidence detector is made from tiles of plastic scintillator which are read out through wavelength-shifting fibers and miniature phototubes. This segmented structure reduces self-vetoing due to backscattering from the tracker/convertor.

The telescope has a modular design with 16 individual tracker/converter modules in a 4×4 array of identical towers. Each tower is semi-independent with its own tracker and calorimeter; its height is 84 cm and lateral dimensions 40 cm $\times$ 40 cm. Unlike EGRET, there will be no consumables to limit the mission which is conservatively planned to have a lifetime of five years.

This next-generation telescope, which will be launced on a NASA rocket,

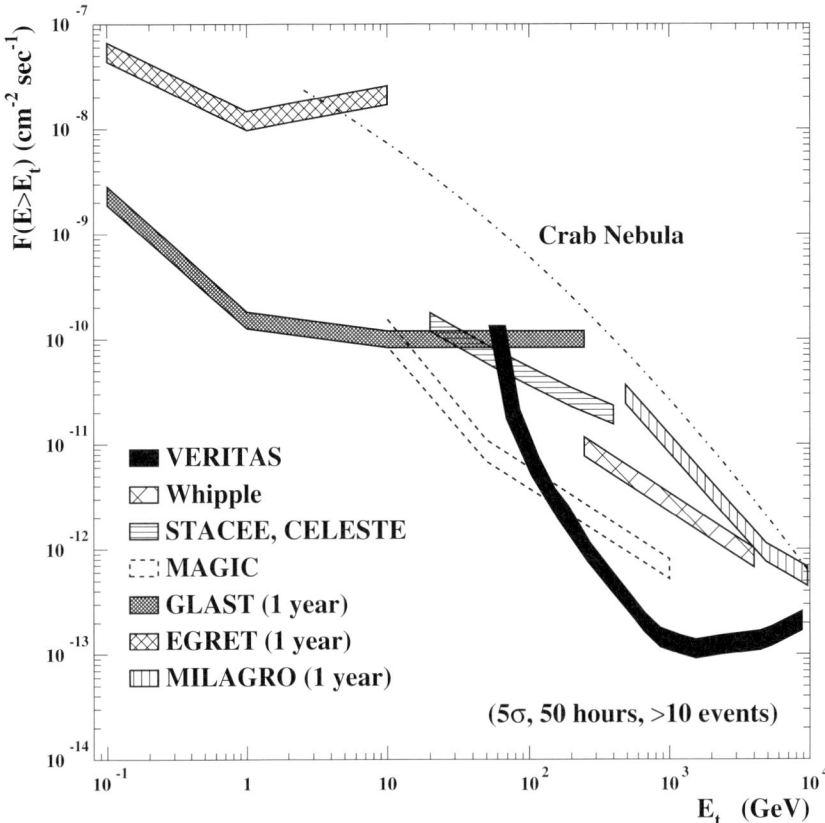

Figure 3.7. Comparison of the point-source sensitivity of various existing and proposed space and ground-based telescopes: Whipple, MAGIC, VERITAS/HESS, CELESTE/STACEE; GLAST, EGRET and Milagro. The sensitivity of MAGIC is based on the availability of new technologies, e.g. high quantum efficiency PMTs, not assumed in the other experiments. EGRET, GLAST, and Milagro are wide field instruments and, therefore, ideally suited for all sky surveys.

will operate in the range 20 MeV–300 GeV, with a scheduled launch date of 2006. GLAST will surpass EGRET by a factor of 10–40 in most parameters (figure 3.6). A comparison of the two missions is given in table 3.1.

Remarkably, this solid state technology can achieve its dramatic improvement over EGRET, outlined in table 3.1, with an instrument that will only be twice as heavy (3000 kg). A comparison of the flux sensitivity of GLAST with EGRET and past, present, and future ground-based experiments is shown in figure 3.7.

> **Historical note: CGRO rescue**
>
> It was the US Federal budget-crunch time in the late 1970s and the NASA budget, as usual, was in jeopardy. The decision of whether NASA should be authorized to proceed with the Gamma Ray Observatory mission (later to be called the Compton Gamma Ray Observatory (CGRO)) came to the desk of President Jimmy Carter. The previous night President Carter, who had trained as a physicist, had read a book on black holes. He asked his aides if the Gamma Ray Observatory would be used in the study of black holes. On being assured that it would, he authorized the 'new start' and the Gamma Ray Observatory became the second of NASA's Great Observatories (the others being the Hubble (optical) Telescope, the Chandra (x-ray) Telescope and the yet to be launched SIRTF (infrared) Telescope). The Challenger shuttle disaster was to set back the CGRO launch date a number of years but it was eventually launched in April 1991.
>
> During the release of CGRO from the shuttle bay the antenna arm refused to deploy, a potential disaster for the mission. It required a space walk by astronaut, Jay Apt, to force the recalcitrant antenna arm to release. In the eyes of many scientists that simple manoeuvre justified all the hassle associated with the launch of scientific missions from manned shuttles.
>
> The anecdote concerning President Carter was told to me by the late Walter Sullivan, for many years the doyen of *New York Times* science writers. It was his recently published book, *Black Holes* [7], that the President had been reading. Walter Sullivan hence felt a personal responsibility for CGRO!

References

[1] Ahlen S P *et al* 1994 *Nucl. Instrum. Methods* A **350** 351
[2] Barbiellini G *et al* 2001 *Proc. 'Gamma-Ray Astrophysics 2001' (Baltimore) (AIP Conf. Series 587)* ed S Ritz, N Gehrels and C R Schrader (New York: AIP) p 774
[3] Barthelmy S D 2001 *Proc. 'Gamma-Ray Astrophysics 2001' (Baltimore) (AIP Conf. Series 587)* ed S Ritz, N Gehrels and C R Schrader (New York: AIP) p 781
[4] Fichtel C E and Trombka J I 1997 *Gamma-Ray Astrophysics: New Insight into the Universe (NASA Reference Publication 1386)*
[5] Gehrels N and Michelson P 1999 *Astropart. Phys.* **11** 277
[6] Schonfelder V 2001 *Proc. 'Gamma-Ray Astrophysics 2001' (Baltimore) (AIP Conf. Series 587)* ed S Ritz, N Gehrels and C R Schrader (New York: AIP) p 809
[7] Sullivan W 1979 *Black Holes; the Edge of Space, the End of Time* (New York: Barnes and Noble)
[8] Weekes T C 2001 Gamma ray telescopes *Encyclopedia of Astronomy and Astrophysics* (Bristol: Institute of Physics Publishing)

Chapter 4

Galactic plane

4.1 Study of the galactic plane

The strongest, and hence first, gamma-ray source to be detected (see historical note: first light), the galactic plane remains one of the prime objects for study in gamma-ray astrophysics and one of its major contributions to astrophysics. That this should be so is attributable to the great difficulty that the general study of the plane presents (because of our immersion in it) and one of the great properties of cosmic gamma rays, their ability to penetrate interstellar matter. The study of the diffuse component of the gamma-ray flux from the Galaxy provides a powerful tool to study the cosmic radiation in the Galaxy as a whole. The galactic plane is unique in that it is the only cosmic source thus far detected where one can unequivocally identify hadrons as the progenitors for at least part of the observed spectrum. However, in a sense the galactic plane is not a source but a medium through which cosmic rays propagate and, hence, the true source of hadronic cosmic rays remains elusive.

It is our position in the midst of the galactic plane that complicates its study; we are too close for a comfortable and objective examination. It is as if we are deep in a dense forest (the Galaxy) where we have a clear view of the nearest trees (the stars) but our vision of the forest as a whole is obscured and we cannot even discern the patterns in the distribution of the trees except in our immediate vicinity. The boundaries of the Galaxy are obscured not by the trees but by the underbrush (interstellar dust). Most of our knowledge of the Galaxy comes, not from its direct study, but by analogy with distant galaxies to which we assume our own to be similar. However, while we can find some verification for the spiral structure in the nearby stellar patterns, it is doubtful that we could have deduced these properties based on studies of our immediate stars alone.

The matter in the Galaxy is composed of stars, interstellar gas, dust, and cosmic rays; there may also be dark matter of unknown composition. By analogy with the Andromeda Nebula, our nearest large galaxy and, to some extent, our twin galaxy, there are 10^{11} stars in the Galaxy whose total mass is of order

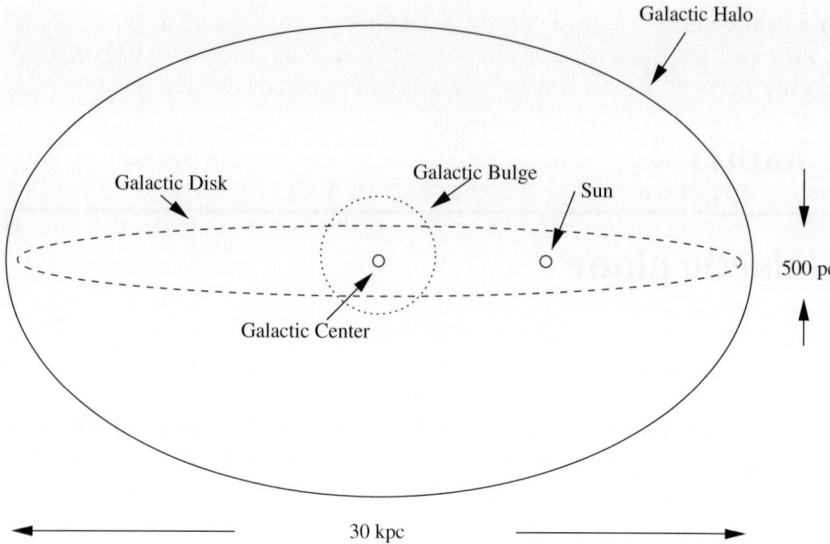

Figure 4.1. Cross section of the Galaxy in the plane perpendicular to the galactic plane showing approximate dimensions.

10^{12} $M_\odot$. The interstellar gas (atomic and molecular) constitutes a total mass of 10^9 $M_\odot$.

A simple model of the Galaxy is that the mass is concentrated mostly in a thin disk of thickness 0.5 kpc and radius 15 kpc. The Solar System is 8.5 kpc from the galactic center (figure 4.1).

Optical astronomy, which has formed the foundation of knowledge of so many astronomical phenomena, fails completely in the study of the Galaxy. Interstellar dust, a relatively minor component of the Galaxy, forms an impenetrable barrier for visible photons. An optical photon has an optical depth in the galactic plane of only 0.1 kpc and, hence, the galactic center is totally obscured. In contrast, the absorption of gamma rays is dependent only on the total grammage traversed so that a gamma-ray photon can traverse the entire length of the Galaxy virtually without attenuation (contrast this with the contrary situation in the earth's atmosphere!).

Similarly, interstellar gas absorbs x-rays almost completely so that x-ray astronomy can tell us little about the structure of the Galaxy. Radio waves suffer virtually no absorption in the plane and they, together with gamma rays, provide the best means of exploring the galactic forest.

The dynamics of the Galaxy are not fully understood. It is observed that the energy density of the cosmic radiation, of the ambient magnetic field, of the turbulence in the interstellar gas, and of the average density of visible radiation are all of the same order, ~ 1 eV cm^{-3}. To what extent this is a coincidence or a demonstration of the equipartition of energy is unknown. In general, these components must be seen as disruptive with tendencies to force the Galaxy apart: to balance these forces gravity acts as a counteracting force and ensures the stability of the Galaxy as a whole [7].

In the study of the Galaxy, we make three fundamental assumptions which seem reasonable but must be recognized as assumptions:

- The Galaxy is like others that we can observe in greater detail and on larger scales, i.e. the Milky Way is a large galaxy with well-defined spiral features (spiral type Sb) and it is not pathological in any way. This seems a reasonable assumption and is partially verified by the observation of spiral features in the distribution of the 21 cm radio line and of pulsars.
- The conditions that we observe in the immediate vicinity of the Solar System are typical of the Galaxy as a whole and in particular, the density and composition of the cosmic radiation, is not unique to our immediate neighborhood. This assumption is less well justified since we know that the cosmic ray distribution is influenced by local circumstances such as supernova explosions, molecular clouds, etc. It is possible that we live in a local bubble and that our cosmic ray densities, particularly the cosmic electron densities, are influenced by events in the recent past.
- The cosmic radiation is a galactic phenomenon with its sources and acceleration processes within the Galaxy. This is the canonical view up to energies of 10^{14} eV and there is evidence in favor of this view from gamma-ray observations of the Small Magellanic Cloud (see chapter 10). However, at extremely high energies ($>10^{18}$ eV) it is almost certainly not the case and in the intermediate ultra high energy range (10^{14}–10^{18} eV) the jury is still out.

The physics of the processes that are thought to produce the observed galactic gamma radiation are well understood:

- cosmic electrons radiating gamma rays by bremmstrahlung with interstellar gas—probably most important at low energies;
- cosmic ray hadrons producing π's in collisions with interstellar gas, the π^0's decay to gamma rays with a characteristic spectral feature—most important at intermediate energies; and
- cosmic electrons inverse Compton scattering on soft photons to give gamma rays—probably most important at high energies.

Although the physics is well known (see appendix), the actual values and distributions of the components in the interactions are not and it is this that is the challenge in interpreting the strong galactic plane signal.

4.2 Gamma-ray observations

4.2.1 HE observations

The galactic plane was first observed by OSO-3 (see historical note: first light). The distribution was clearly structured in latitude and longitude with the intensity peaked in the direction of the galactic center. Subsequent observations by the more sophisticated spark chamber telescopes of SAS-2 and COS-B mapped these distributions in greater detail and permitted the identification of several point sources. The distribution of gamma rays was generally found to correlate well with known features of the Galaxy such as the spiral arms. The factor of 20 improvement in sensitivity offered by EGRET, combined with its longer exposure, produced the most detailed maps and permitted serious comparisons to be made of the gamma-ray galactic plane with that predicted from models [8].

The detailed analysis by the EGRET group was based on Phase I and II of the EGRET mission (approximately the first 28 months of data) when the telescope had optimum and uniform sensitivity. The standard EGRET data-processing was used to map the distribution of photons as a function of energy into bins $0.5° \times 0.5°$ (galactic latitude and longitude). Eleven energy intervals were used (from 30 MeV to 30 GeV). Since EGRET had essentially no background, the photon distribution could be presumed to come from a combination of the galactic plane diffuse flux, known discrete sources, unresolved discrete sources, and the extragalactic background flux. The known discrete sources, although individually strong, contribute only 9% of the total flux above 100 MeV; the contribution from unresolved discrete sources is assumed to be small and is usually neglected in the simplest models.

4.2.2 VHE observations

Atmospheric Cherenkov telescopes (ACTs) are not ideally suited for the study of extended sources because of their limited fields of view. Air shower arrays are more suited to the study of such sources, e.g. the galactic plane, but because the energy thresholds of the arrays are relatively high, the sensitivity is limited. To date neither technique has been successful in measuring a gamma-ray flux at energies in excess of 100 GeV.

Two ACT groups (Whipple and HEGRA) have attempted to measure the gamma-ray flux from the galactic plane at longitude ~40° at VHE energies. Although the observations were of limited duration and showed no evidence for an excess using a variety of assumptions about the width of the plane, they are clearly in conflict with a simple extrapolation of the flux from the EGRET measurements (figure 4.2).

Upper limits to the ratio of the gamma-ray flux to the cosmic ray flux have also been reported from air shower experiments at higher energies [11]— a representative set of limits are given in table 4.1. These limits correspond to the inner galactic plane as seen from the Northern Hemisphere which is assumed to

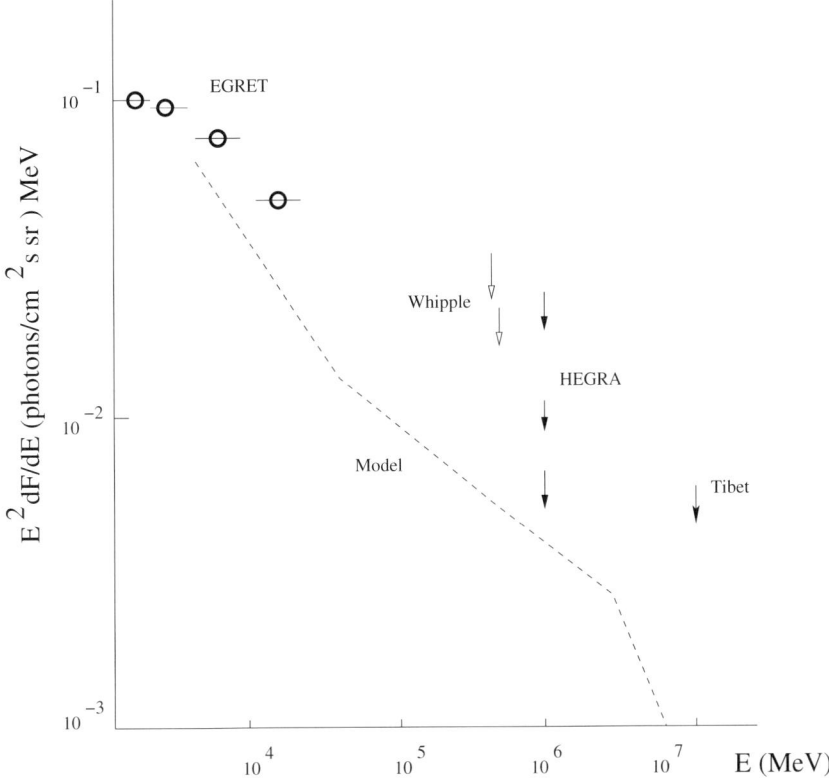

Figure 4.2. Upper limits to the diffuse gamma-ray flux at GeV–TeV energies from the EGRET, Whipple, HEGRA (numbered points) and Tibet experiments. The data correspond to observations at galactic longitude 35° to 45° and latitudes ±2° [10]. The broken curve represents the prediction of the 'leaky box' model [3].

Table 4.1. Upper limits from array experiments.

Experiment	Energy threshold (TeV)	Ratio $(I_\gamma/I_{cr} \times 10^{-5})$	Ref.
Tibet	10	<60	[2]
EAS-TOP	130	<40	[1]
CASA	180	<2.6	[4]

be ±5° wide in galactic latitude. These limits are tantilizing close to the predicted fluxes in some simple models.

4.3 Interpretation

In the galactic plane [8], the calculation of the yields at HE energies from the three gamma-ray production processes listed earlier requires a knowledge of the following:

- the composition of the cosmic radiation hadron component which is not pure protons but is ~10% helium and ~1% heavier nuclei,
- the composition of the interstellar gas,
- the distribution of the interstellar gas,
- the ratio of cosmic electrons to protons of the same energy—this is about 1:100 near the Solar System but may vary across the Galaxy, and
- the distribution of soft photons, e.g. in the optical band, from stellar and dust emission (required to evaluate the inverse Compton component).

In principle, the gamma-ray flux could be calculated exactly if all these quantities were known independently since the physics is well known and no other processes are thought to make serious contributions [14]. In practice, these quantities are not known exactly and we must use an *a posteriori* model to constrain them by comparison with the gamma-ray observations. Thus, we extend our knowledge of the galactic plane somewhat and our knowledge of its components.

The interstellar gas is largely hydrogen but can come in one of three forms: atomic (HI), molecular (H_2), or ionic (HII). In practice, the ionic component is small ($\sim 10^{-3}$ atom cm^{-3}) and is often neglected. Considerable effort has gone into estimating of the other two which have similar total masses but significantly different distributions.

The density of atomic hydrogen can be measured using the hyperfine transition of hydrogen which is detected by radio telescopes as a line at a wavelength of 21 cm. This is a major field of study in radio astronomy and the techniques have been refined to give detailed maps of 21 cm emission and absorption in the Galaxy (and in other, more distant, galaxies). With some confidence, these velocity maps of the Doppler-shifted line can be transformed into maps of hydrogen density. These are believed to be effective tracers of the spiral arms. The typical density is 1 atom cm^{-3}.

The molecular component cannot be measured directly. However, the carbon monoxide line (^{12}CO) at 2.6 mm can be easily seen; if the simplistic assumption is made that the CO and H_2 distributions are similar, then the molecular distribution can be estimated. The relative density of the two molecules is unknown but is initially taken to be the same as that observed near the Solar System. The exact value is treated as a free parameter. A value of the H_2/CO ratio in the range 1.5–2.0 × 10^{20} seems to fit the observations. The molecular distribution is uneven and associated with 'molecular clouds', the largest substructures in the Galaxy. Within these the density can be as high as 10^4 atom cm^{-3}.

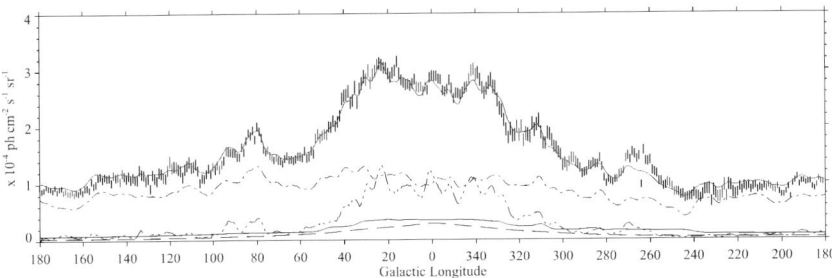

Figure 4.3. Galactic distribution in latitude of gamma rays ($E > 100$ MeV) as measured by EGRET. This cut is made at longitude $= -210°$ [8]. (Figure: S Hunter.)

The simplest model is one that assumes that the Galaxy is filled uniformly with cosmic rays with the same density that is found in the vicinity of the Solar System and that the distribution is dominated by proton–proton interactions. The structure in the gamma-ray maps is then supposed to arise from differences in the density of interstellar matter in the plane. When confronted with observations, this model fails completely and is replaced by one in which the cosmic ray density varies across the Galaxy and is correlated with the matter density. The degree of correlation, i.e. the assumed distance scale of the correlation, is another free parameter and there is no independent estimate of what it should be. It is taken to be ~1.75 kpc and independent of energy.

There is one other quantity that must be supplied—the intensity of the diffuse extragalactic background at 100 MeV. This is extremely difficult to measure or to estimate on theoretical grounds (chapter 14). However, it is small compared with the galactic plane contribution. From the EGRET measurements it can be represented by a power law with spectral index -2.1 and integral flux above 100 MeV of 1.17×10^{-5} photons cm^{-2} s^{-1} sr^{-1}.

Thus, the galactic plane distribution in latitude and longitude observed by EGRET can be compared with a model of the galactic plane with the three free parameters. The prominent discrete sources can be subtracted and the result is a fairly successful fit (figures 4.3 and 4.4) over a range of energy bands. This gives some confidence that the model is realistic, that the three processes postulated to dominate are, in fact, in operation, and that the progenitors are a mixture of hadrons and electrons.

The broader source contribution from the outer Galaxy (galactic anti-center), compared with the inner Galaxy (galactic center), can be understood in terms of geometry. The close proximity of the anti-center region of the plane gives a broad source whereas the opposite direction is dominated by the more distant galactic center where the plane, being further away, appears narrower (figure 4.1). A more detailed examination shows correlation also with some nearby large molecular clouds (Orion and ρ Oph).

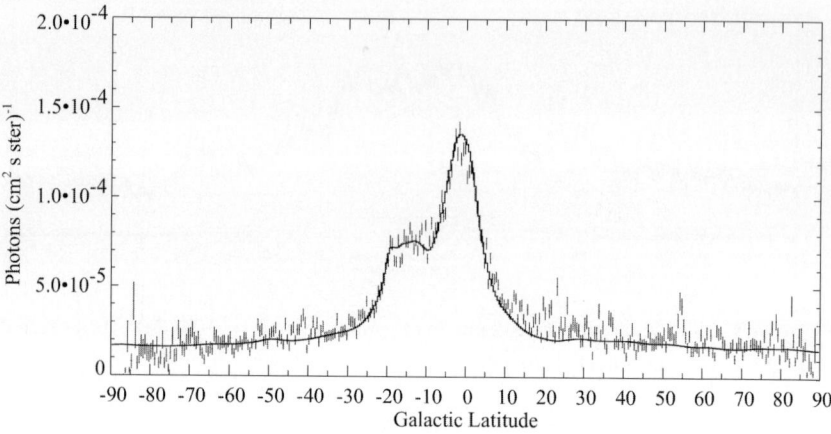

Figure 4.4. Galactic distribution in longitude of gamma rays ($E > 100$ MeV) in the latitude interval $\pm 10°$. The EGRET measurements are shown as data points and the full curve is the fitted distribution which is the sum of four components: (i) dash-dot, cosmic ray (CR) + HI interactions; (ii) dash-triple dot, CR + H_2 interactions; (iii) full curve (lower), CR + HII interactions; and (iv) dash-dash, inverse Compton component [8]. (Figure: S Hunter.)

4.4 Energy spectrum

The validity of the HE galactic plane model is supported by comparison of the observed energy spectrum from the galactic center region with that predicted (figure 4.5) [9]; there is seen to be good agreement between 50 MeV and 1 GeV. At energies below 100 MeV the observed spectrum is probably dominated by gamma rays produced by electron bremmstrahlung on the interstellar gas. Of particular note here is the prominent bump in the spectrum near 100 MeV. This is associated with the rest mass of the π^0 and was predicted to be a feature of most gamma-ray source spectra. The identification of this feature means that the progenitor particles must be almost certainly hadrons.

At medium latitudes (2°–10°), the observed spectrum appears generally uniform with longitude. Above 1 GeV there is evidence that the spectrum at low latitudes (<2°) is softer from the outer Galaxy than from the inner Galaxy (which is similar to that at medium latitudes). This contradicts an earlier finding from the COS-B mission and has not been explained.

The agreement at lower and higher energies is not satisfactory and indicates that the simple model requires modification. At low energies where the emission is thought to come predominantly from electron bremsstrahlung, the observed spectrum was measured by COMPTEL and the intensity is greater than predicted. The excess could be the result of unresolved discrete source contributions (the angular resolution is significantly worse at lower energies) or because there is

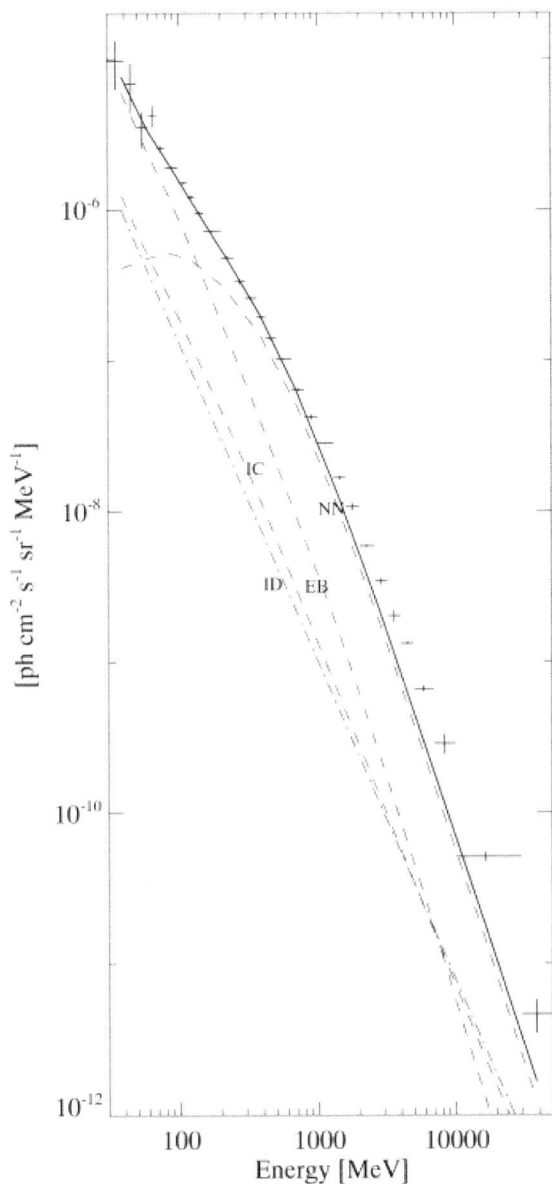

Figure 4.5. The energy spectrum of the diffuse component of gamma rays as measured by EGRET towards the galactic center (longitude 300°–60°, latitude ±10°). The data points are the EGRET measurements and the bold full curve is the sum of all the components using the best-fit model. NN = nucleon–nucleon collisions. EB = electron bremsstrahlung. IC = inverse Compton; and ID = isotropic diffuse flux (extragalactic) [9]. (Reproduced with permission from the *Astrophysical Journal*.)

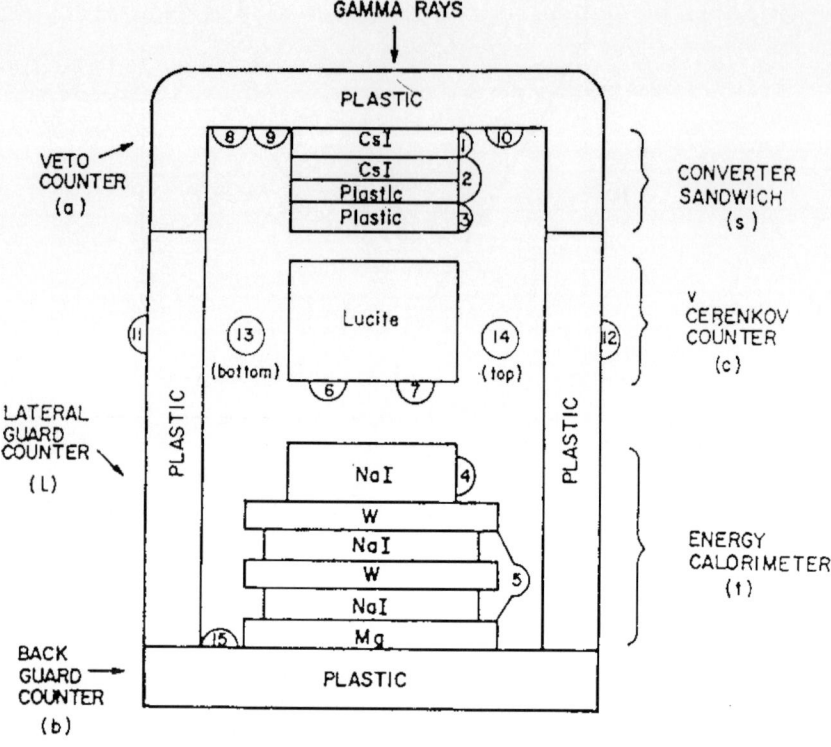

Figure 4.6. Schematic diagram of the OSO 3 gamma-ray detector flown in 1967 [6].

a greater density of cosmic electrons in the Galaxy generally than is seen in the vicinity of the earth (but it cannot be too great or it would exceed the output power expected at these energies from supernova remnants (SNRs) and OB stars) [13].

At high energies (>1 GeV), where the measurements all come from EGRET, the difference between the measured and calculated spectrum is more striking and there is no satisfactory explanation for the discrepancy. Because of the reasonably good angular resolution above 1 GeV, it seems unlikely that there is a major contribution from unresolved point sources. Among the reasons postulated by the EGRET team for this discrepancy are that the observed cosmic ray hadron intensity at the earth is not typical of the Galaxy as a whole, that the calculation of the proton–proton interaction has features that are not included in the calculation, that there is an error in the calibration of the EGRET instrument, or that there are more hard unresolved point sources than anticipated [9].

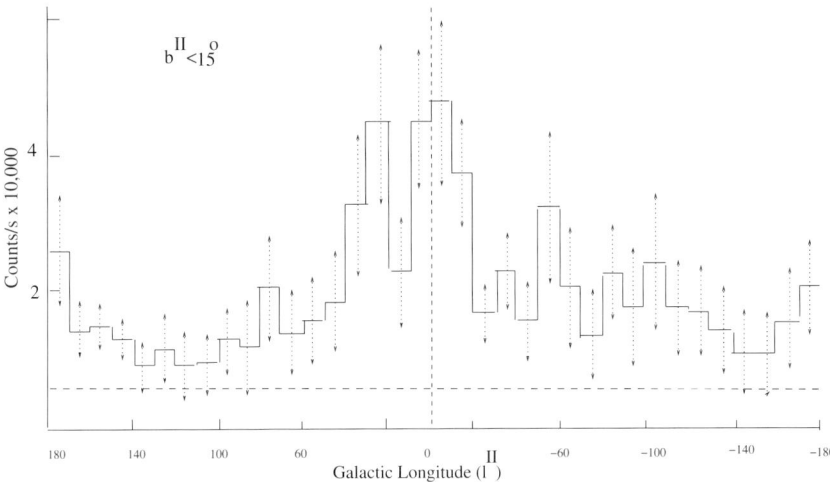

Figure 4.7. Galactic distribution in longitude of gamma rays as measured by OSO 3 for latitudes ±15° [6].

Historical note

The galactic plane was established as a HE gamma-ray source by observations with the OSO 3 satellite telescope in 1968 [6]. This was a similar instrument to that used by the same group on Explorer XI to establish the existence of extra terrestrial gamma rays. The scintillation-Cherenkov telescope (figure 4.6) had an area ~ 100 cm^2 and an angular resolution of $\pm 15°$. The detection was consistent with a line source along the galactic plane (figure 4.7) with enhanced emission around the galactic center. Although the angular resolution was not sufficient to resolve the width of the plane, it was apparent that the contribution from the galactic center was broader than the resolution of the instrument. The intensity from the broad region around the galactic center was 3×10^{-4} photons cm^{-2} s^{-1} rad^{-1}. This was the first unambiguous detection of a gamma-ray source.

To question the assumption that the observed hadronic cosmic ray spectrum in the vicinity of the Solar System is representative of the Galaxy as a whole is to question a fundamental assumption about the cosmic radiation in the Galaxy. It is easier to postulate that the electron spectrum is harder in general than that measured locally, in which case the hard inverse Compton component of the gamma radiation above 1 GeV will be increased and the observed spectrum can be accounted for [12]. However, this hardened electron spectrum cannot extend beyond 10 TeV or the upper limits from VHE observations would be in conflict. It is also possible that the spectrum of cosmic rays from SNRs is not uniform (which is plausible) so that the averaged cosmic ray spectrum in the Galaxy is not a smooth power law but exhibits a definite curvature [5]. This could produce the observed gamma-ray spectrum. However, the VHE limits again constrain the acceleration within the SNRs to energies below 100 TeV and present a problem for cosmic ray origin theory.

None of these explanations is completely satisfactory; it may be that the agreement of the spatial distributions of the galactic model is somewhat fortuitous and that the galactic gamma-ray distribution is more complicated than previously considered.

References

[1] Aglietta M *et al* 1992 *Astrophys. J.* **397** 148
[2] Amenomori M *et al* 1997 *Proc. 25th, ICRC (Durban)* **3** 117
[3] Berezhko E G and Volk H J 2000 *Astrophys. J.* **540** 923
[4] Borione A *et al* 1998 *Astrophys. J.* **493** 175
[5] Busching I, Pohl M and Schlickeiser R 2001 *Astron. Astrophys.* **377** 1056
[6] Clark G, Garmire G P and Kraushaar W L 1968 *Astrophys. J. Lett.* **153** L203
[7] Fichtel C E and Trombka J I 1997 *Gamma-Ray Astrophysics: New Insight Into the Universe (NASA Reference Publication 1386)* 2nd edn
[8] Hunter S D *et al* 1994 *Astrophys. J.* **436** 216
[9] Hunter S D *et al* 1997 *Astrophys. J.* **481** 205
[10] Lampeitl H *et al* 2001 *Proc. 27th ICRC (Hamburg)* **6** 2348
[11] Ong R A 1998 *Phys. Rep.* **305** 93
[12] Pohl M 2001 *Proc. 27th ICRC (Hamburg)* ed K H Kambert, G Heinzelmann and C Spiering (University of Hamburg)
[13] Strong A W, Moskalenko I V and Reimer O 1999 *Proc. 5th Compton Symposium (Portsmouth) (AIP Conf. Proc. 510)* ed M L McConnell and J M Ryan (New York: AIP) p 283
[14] Strong A W and Moskalenko I V 2001 *The Universe in Gamma Rays* ed V Schonfelder (Berlin: Springer) p 207

Chapter 5

Supernovae and supernova remnants

5.1 Supernova explosions

The catastrophic explosion of a star is known as a supernova. Although there seem to be a number of ways in which the explosion can occur, the results are similar. A star explodes with the emission of some 10^{51} erg of energy in a very short time interval. The star is destroyed although its remains can be detected for thousands of years thereafter over a wide range of wavelengths. Supernovae play a critical role in high energy astrophysics for many reasons, among them the production of heavy elements, the formation of new stars, and the acceleration of cosmic rays. They have also been used as standard candles for measurements of distance on the cosmological scale.

Supernova explosions have been high on the list of potential sources of gamma rays from the earliest experiments. On the stellar scale they are the most energetic objects known, with radiation detected across a wide range of wavelengths. The only object outside the Solar System to be detected by its elementary particle emission (neutrinos) was the supernova in the Large Magellanic Cloud, SN1987A (see historical note: SN1987A). Gamma rays that result from supernova explosions may be detectable either in the first few seconds of the explosion as a gamma-ray burst (chapter 13), as the steady, but periodic, emission from the pulsar, as the rotating core of the exploded star, or as the expanding outer shell of the star known as the supernova remnant (SNR).

Because of their brightness, supernovae can be studied in detail over many wavebands. Despite the wealth of observations, the exact mechanism at work in the explosions is not fully understood. Supernovae are classified mostly based on their optical emissions. The two most important types are known as type Ia and type II.

Type Ia occur after the formation of a white dwarf in a binary system; accretion from the companion causes the star to overload and undergo a thermonuclear explosion. No hydrogen emission lines are seen in the optical spectra and although they occur in all types of galaxies, type Ia are not observed in star-forming regions. All of the stellar mass is ejected and no core remnant is observed. Types Ib and Ic are more akin to type II. Examples of type Ia are SN1006, SN1572 (Tycho), and SN1604(Kepler).

Type II are associated with the rapid evolution of massive stars (with the ejection of some 1–10 $M_\odot$ of stellar material at near relativistic velocities). When the core of the star collapses there is an implosion which leaves behind a neutron star (which often becomes a pulsar). The rebound from the formation of the compact object generates a shock wave which propagates out through the outer layers of the star. At one time it was thought that this shock wave would be sufficient to form the expanding shell. Current theory finds that this is not sufficient and that the energy transport comes via a dense wave of neutrinos emitted from the neutron star. The optical emission spectra from these outbursts show a normal abundance of elements including hydrogen. These explosions are observed in star-forming regions such as in the arms of spiral galaxies. The Crab Nebula and SN1987A are examples of type II supernovae.

Remarkably, despite their quite different origin, both types of supernovae have similar energy of ejection, of some 10^{51} erg.

5.2 Energy considerations

The hypothesis that supernovae are the most likely source of cosmic rays within the Galaxy (at least up to energies of 100 TeV) is arrived at by simple energy considerations [4]. It is generally assumed that the observed cosmic ray density fills not only the Galaxy but the galactic halo as well, a total volume of some $1–5 \times 10^{68}$ cm^3. The spectral distribution of cosmic ray energies is soft so that most of the energy is concentrated in the lower energies, a density of about 10^{-12} erg cm^{-3} or 0.5 eV cm^{-3} above 1 GeV. The total energy is, therefore, $\sim 1–5 \times 10^{56}$ erg. This number is relatively non-controversial. It is more difficult to calculate the loss (and, hence, the necessary replenishment rate) if the cosmic ray density is in quasi-equilibrium. One method is to use the relative abundance of isotopes produced in the cosmic radiation by spallation, i.e. the breakup of heavier nuclei in collisions. The half-life of the beryllium isotope, Be^{10} is 1.5×10^6 years. The measurement of its concentration relative to the other beryllium isotope in the cosmic radiation can give a measure of the average lifetime of a cosmic ray in the Galaxy; this yields a value of about 3×10^7 y. Other estimates [7] give a fairly consistent value of 1.4×10^7 y, leading to a replenishment rate $\sim 2 \times 10^{41}$ erg s^{-1}. This is the minimum value generally used to characterize the fundamental problem of cosmic rays: what is the source within the Galaxy that can provide this rate of emission of hadronic cosmic rays for most of the lifetime of the Galaxy?

A variety of possible sources within the Galaxy on the basis of their emitted power (without consideration of how the acceleration might be achieved) are summarized in table 5.1 [4]. In most cases, either the maximum energy emitted

Table 5.1. Potential sources of galactic cosmic rays.

Sources	Example	Power/source (erg s^{-1}) (maximum)	Number of sources/ frequency (maximum)	Total power (erg s^{-1}) (maximum)
Normal stars	Sun	10^{24} (steady)	10^{11}	10^{35}
Magnetic stars	53 Cam	10^{30} (steady)	10^{9}	10^{39}
Active stars	T Tau, UV Cet	10^{34} (steady)	5×10^{4}	5×10^{38}
Nova	Nova Serpentis 1970	10^{45-46} (explosion)	3×10^{-6} s^{-1}	$3 \times 10^{39-40}$
Supernova	Cas A, Crab	10^{50} (explosion)	10^{-9} s^{-1}	10^{41}

or the frequency of the phenomenon is insufficient. Nova are almost sufficient but there is no independent evidence of high energy particle production in these stellar outbursts to bolster their case. However, SNRs are clearly the sites of expanding nebula with gas expanding with velocities of up to $10\,000$ km s^{-1}. They are also reservoirs of relativistic particles; originally it was thought that these particles were the direct result of the explosion but current thinking is that they are accelerated in a two-stage process.

Assuming that the observed density of cosmic radiation observed in the Solar System is typical of the Galaxy as a whole and that this has not changed significantly over the lifetime of the Galaxy, there is simply no other energy source within the Galaxy capable of maintaining this density other than on-going supernova explosions.

The energy that might be emitted in cosmic rays from a supernova explosion is arrived at by making a reasonable estimate of the total energy emitted (from observations) and the fraction of such energy that might be in hadronic cosmic rays. This is complicated by the fact that there are many different types of supernova explosion with different total energies and frequencies. However, there is no doubt that they are the most powerful phenomenon known in the Galaxy and a total energy output of 10^{52} erg is strongly suggested. A conversion efficiency of 1% to cosmic rays is not excessive. A frequency of one explosion per 30 years is arrived at by observations of supernova explosions in similar galaxies to our own. A total input of 10^{40}–10^{41} erg s^{-1} is, thus, reasonable. Supernova blast shocks are the only galactic source capable of satisfying the energy required for the total production of galactic cosmic rays as well as exhibiting evidence for particle acceleration. The caveat is that these must have a high efficiency for converting the kinetic energy of supernova explosions into high energy particles.

The discussion thus far has concentrated on the total power in the cosmic radiation, not its distribution with energy. Only supernovae provide the necessary total power to be the sources of the 'low' energy cosmic radiation in the Galaxy. However, this is only part of the problem: there is a wide range of cosmic ray energies observed (see table 5.2). In fact, cosmic ray particles have been observed up to 3×10^{20} eV. It is impossible to accelerate such particles in the limited dimensions of a stellar explosion [5]. The canonical wisdom is that supernova explosions can only account for the observed cosmic radiation up to energies of 100 TeV and above this one must look to other sources, probably outside the Galaxy.

5.3 Acceleration

The most likely mechanism by which cosmic rays are accelerated in supernova outbursts is in the shock waves caused by the outburst that may persist for thousands of years after the explosion. There is little doubt that shock waves are produced by the initial implosion/explosion and that they propagate into

Table 5.2. Cosmic ray energy densities.

Energy	Density (eV cm^{-3})	Origin
>1 GeV	0.5	Galaxy
>100 TeV	10^{-9}	Halo
>10 PeV	10^{-13}	Extragalactic

interstellar space where they can be observed as the expanding gas shells of SNRs. Shock acceleration of relativistic particles is a favorite mechanism for theoretical speculation since it is relatively well understood and easy to calculate. Also there is observational support for the acceleration of particles to super-thermal energies in interplanetary space. There is, however, no direct evidence for shock acceleration in operation at truly relativistic energies and one of the hopes for gamma-ray astronomy is that it will provide the first direct observational evidence of shock acceleration of hadrons up to energies of 100 TeV. There is evidence that this is observed for electrons but there is, as yet, not definitive observations of hadron shock acceleration.

The supernova explosion ejects the outer layers of the star which propagate into the interstellar medium producing a shock wave. As it moves out, it is resisted by the interstellar medium. For a typical interstellar density of 1 proton cm^{-3}, it will take about 10^3 years for 10 M$_\odot$ of material moving with a velocity of 5000 km s^{-1} to sweep up its own mass of interstellar material [3]; this is the characteristic time of the supernova expansion and the time when most of the acceleration occurs. For first-order Fermi acceleration and the previous conditions, the maximum particle energy that can be achieved for a magnetic field in the interstellar medium of 3 μG is less than $Z \times 30$ TeV [6].

The model of diffusive shock acceleration, which provides a plausible mechanism for efficiently converting the explosion energy into accelerated particles, naturally produces a power-law spectrum of $dN/dE \propto E^{-2.0}$. This is consistent with the inferred spectral index at the source when the observed local cosmic ray spectrum is $dN/dE \propto E^{-2.7}$, after correcting for the effects of propagation in the Galaxy (which causes a steepening in index of ~ -0.6).

5.4 Detection at outburst

Since supernova explosions involve a tremendous release of energy in the first few seconds of the outburst, it is natural that they should be considered as prime candidates for emission of detectable fluxes of gamma rays during this time. They are also observed to be sources of low energy gamma rays in the months/years after the explosion. Gamma rays from the radioactive decay of the elements released in the explosion are the dominating factor in the light curve of the supernova in the months following the outburst. One such element is nickel. Ni^{56} decays to Co^{56} with a half-life of 6.1 days and this, in turn, decays with the emission of β and gamma rays with a half-life of 111 days. From 120 to 1800 days after the outburst, the light-curve has been seen to decay with this half-life. In the case of SN1987A, the supernova observed in the Large Magellanic Cloud, the low energy nuclear gamma-ray line emission was observed directly by space telescopes. The first (and only) detection of high energy neutrinos (beyond the sun) was also from this source. This is the closest supernova observed in recent times, i.e. in the last 300 years. The neutrino emission was only detectable for 20 s. Unfortunately, there were no HE gamma-ray telescopes in orbit that could have detected the direct gamma-ray emission from the blast.

However, there was a VHE telescope in operation in the southern hemisphere (the University of Durham telescope at Narrabri, Australia) and attempts were made to observe SN1987A within days of the report of the optical detection; only upper limits were reported. The expanding gas shell is a strong absorber of gamma rays produced inside the shell and even if gamma rays of GeV energy and above are produced, it is unlikely that they can escape to produce a detectable signal until many years after the explosion.

As we shall see in chapter 13, certain types of supernovae may, in fact, be the source of beamed gamma-rays bursts, at least at lower energies. The first eagerly awaited, and long overdue, supernova in the Galaxy since the development of modern astronomy, will be a prime source of study by gamma-ray astronomers. Ironically, unless it occurs in an optically obscured region of the Galaxy, the optical brightness of the decaying supernova will be too bright for observations by ground-based gamma-ray telescopes using optical techniques! Given the large fraction of the Galaxy that is optically obscured (85%), obscuration is not unlikely. It is probable that the trigger for such observations will come from the next generation of neutrino detectors.

5.5 Supernova remnant classification

There are more than 250 SNRs observed in the Galaxy and they have dramatically different appearances. This is not unexpected; since there is more than one way in which a supernova explosion can occur and the form of the SNR depends strongly on the nearby interstellar medium (e.g. low density or expansion into a neighborhood filled with molecular clouds). Also they have a wide range of ages, distances, and angular sizes. Source confusion or overlap is not unusual. It is not easy to determine the distance or density; the magnetic field, a vital component in any source model, is impossible to measure directly.

The most common type of SNR is characterized by the shell of interstellar material swept up by the expanding shock wave which is clearly visible in x-rays. These are the shell-type SNRs which are observed as rings because of limb-

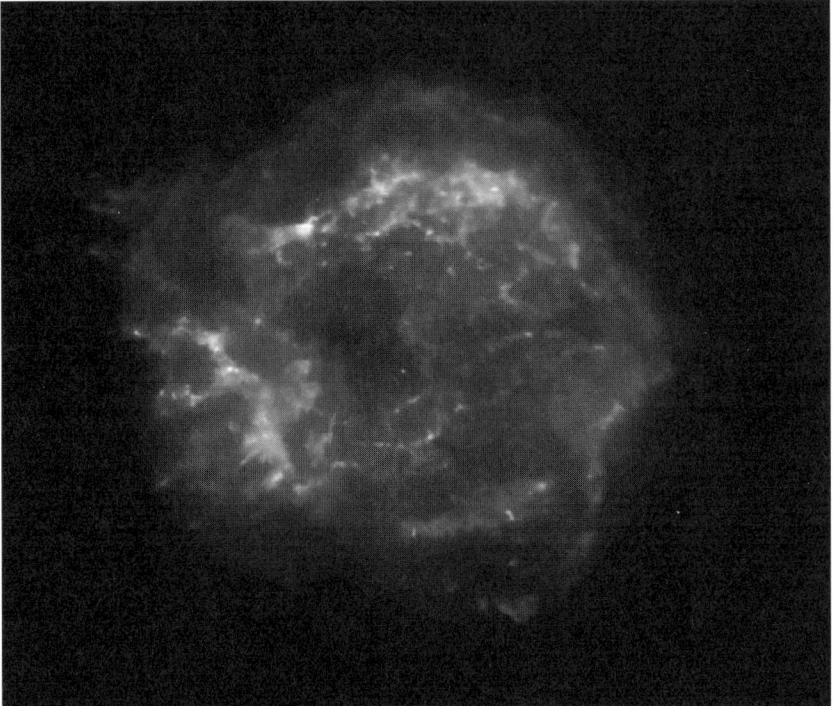

Figure 5.1. Cassiopeia A as seen by the Chandra x-ray telescope. The spot near the center is believed to be the neutron star. (Figure: Chandra/NASA.)

darkening (increased brightness when the spherically symmetric shell is viewed tangentially). Often the ring is distorted due to irregularities encountered in the interstellar medium. Examples of shell SNRs are the well-studied Tycho SNR, Cassiopeia A (figure 5.1) and SN1006.

An SNR, with a pulsar at its center which continually fills the remnant with relativistic electrons, is known as a plerion. The region around the pulsar is dominated by the synchrotron radiation from the relativistic electrons produced by the pulsar; this emission spectrum can range from radio to x-ray and gamma-ray wavelengths. The Crab Nebula, one of the best studied sources in the sky, is the prototype of this kind of object and its properties are relatively well understood (see next chapter).

The distinction between shell-type SNRs and plerions is not sharp. There may be a continuum of SNRs that have both shells and plerions to various degrees. In these intermediate objects, there is a small plerion left behind by the expanding shock wave as well as a detectable shell. The Vela SNR is a prime example of this type with a small plerion (centered on the pulsar) surrounded by a large shell.

X-ray data from the ROSAT and ASCA x-ray telescopes indicate that synchrotron nebula are associated with most of the pulsars that have high spin-down luminosities; such objects must be relatively nearby to be detected. The x-rays point to the presence of relativistic electrons and encourage the search for TeV gamma-ray emission. Observationally, shell-type SNRs are more important for HE gamma-ray astronomy whereas VHE telescopes seem to be sensitive to all types of SNRs.

5.6 SNRs as cosmic ray sources

HE and VHE gamma-ray observations of SNRs may provide the crucial test of the acceleration of hadrons. If the density of the hadronic component of cosmic rays is greater within SNRs than in interstellar space, then the frequency of nuclear collisions will be increased, as will the rate of pion production. Detailed predictions have been made of the resulting gamma-ray production over the full HE and VHE gamma-ray spectrum and the results compared with the current and future sensitivities of gamma-ray telescopes [2]. Surprisingly, the best chances of detection are in the VHE range where the background from the diffuse flux from the galactic plane is not a limiting factor. The gamma-ray fluxes will be enhanced, and in some cases exceeded, by the gamma-ray production from the relativistic electrons, the decay products of the charged pions, via the bremsstrahlung and inverse Compton mechanisms. Hence, the gamma-ray measurements may not be unambiguous but will at least set an upper bound to the hadronic production. In some cases this proves to be a significant constraint on the theory of SNR origin of cosmic rays.

At HE and VHE energies, the spectrum of gamma rays produced from pion decay has a similar power-law spectral index as the progenitor protons. The production rate of the latter has been calculated using the diffusive shock acceleration mechanism. The hadron acceleration and resulting gamma-ray luminosity varies as the SNR expands. Initially, when there is free expansion the luminosity is low. However, when the amount of interstellar matter swept up by the expanding shock wave equals the amount of mass ejected (the so-called Sedov phase) the gamma-ray luminosity peaks and remains constant until the accelerated particles begin to escape from the SNR.

If E_{sn} is the energy released in the supernova explosion and θ is the efficiency of conversion of this energy into cosmic rays, then the energy in the SNR in cosmic rays, $E_{cr} = \theta E_{sn}$, the HE gamma-ray flux is given by

$$F_\gamma(> 100 \text{ MeV}) \approx 4.4 \times 10^{-7} (E_{cr}/10^{51} \text{ erg})((D/1 \text{ kpc})^{-2})$$
$$\times (n/1 \text{ cm}^{-3}) \text{ photons cm}^{-2} \text{ s}^{-1}$$

where D is the distance to the SNR. More generally, if the cosmic ray differential energy spectrum in the SNR is a power law with spectral index -2.1, then the integral gamma-ray flux is [3]:

$$F_\gamma(> E_\gamma) \approx 9 \times 10^{-11} (E/1\text{ TeV})^{-1.1} (E_{\text{cr}}/10^{51}\text{ erg})(D/1\text{ kpc})^{-2}$$
$$\times (n/1\text{ cm}^{-3})\text{ photons cm}^{-2}\text{ s}^{-1}.$$

The Tycho SNR is probably just past its Sedov phase and has an angular size of 4 arc-min. Its parameters are estimated as $D = 2.25 \pm 0.25$ kpc, $n \approx 4$ cm^{-3} and $E_{\text{sn}} \approx 8 \times 10^{50}$ erg [2]. For $\theta = 0.15$, this gives $F_\gamma(>100\text{ MeV}) \approx 4.4 \times 10^{-8}$ photons cm^{-2} s^{-1} and $F_\gamma(> 1\text{ TeV}) \approx 9 \times 10^{-12}$ photons cm^{-2} s^{-1}. These are below the sensitivities of current detectors, e.g. EGRET, Whipple but are well within the capabilities of the next generation of detectors, e.g., GLAST, VERITAS, HESS. Ideally, for detection we require a supernova explosion which produces cosmic rays with high efficiency, which is nearby, which is at, or close to, its Sedov phase (approximately 100–1000 years old), and which occurs in a relatively dense region of interstellar space. At $E_\gamma > 1$ TeV, $n = 0.1$ cm^{-3} and $D = 1$ kpc, the flux can exceed 10^{-11} photons cm^{-2} s^{-1} which is detectable with currently available VHE telescopes. By contrast, the HE fluxes are likely to be comparable with the diffuse flux from the galactic plane which makes their detection much more difficult. The values of D and n, even for nearby SNRs, are not well determined and, hence, these predictions are always somewhat uncertain.

Historical note: SN1987a

The discovery of the supernova in the Large Magellanic Cloud in 1987 by an observer using a small optical telescope in Chile was one of the most exciting serendipitous events in modern astronomy. At a distance of only 50 kpc, this was the first instance in which a supernova explosion could be studied by a whole battery of sophisticated instrumentation. Since its progenitor was a known object, a 12th magnitude star known as Sanduleak-69.202, it represented the first instance in which a spectral star type (a B3 star) could be assigned to the star before it exploded [1]. The fact that the initial explosion was detected in two neutrino experiments was a major boost to theoretical models of explosion mechanisms as was the subsequent detection of the expected gamma-ray lines from radioactive decay of cobalt. The bolometric light curve was monitored before it reached its maximum (80 days after discovery) and for years after the initial explosion. The detection of optical rings (figure 5.2) around the star were evidence of previous ejections of stellar matter into the interstellar medium and possibly of the presence of a companion compact star. They also provided an explanation of why the supernova was rather faint for a type II supernova (absolute magnitude -15.5, instead of the usual -18). The only disappointing aspect of Supernova 1987A is that it has so far failed to reveal the pulsar that might have been formed in the supernova collapse.

76 Supernovae and supernova remnants

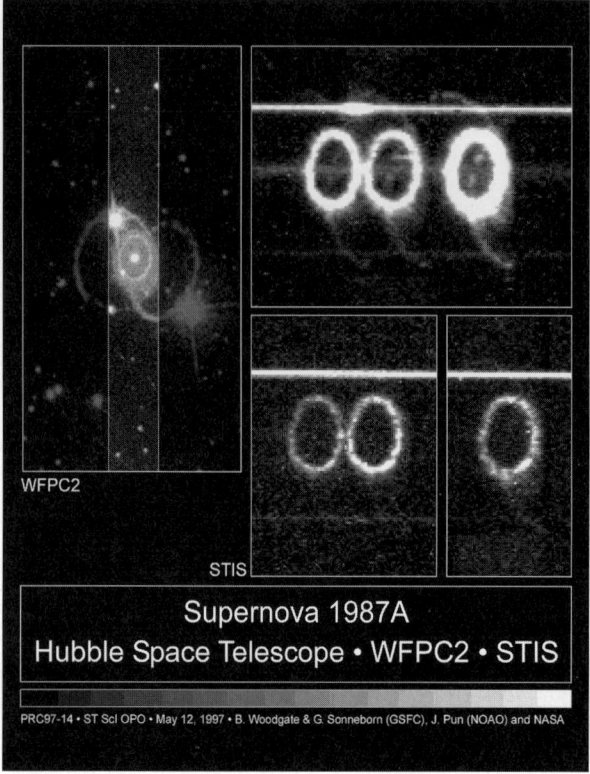

Figure 5.2. Optical images of the three rings around SN1987A which are illuminated by radiation from the supernova. The images were recorded by the WFPC2 and STIS cameras on the Hubble Space Telescope. (Figure: STScI/NASA.)

References

[1] Carroll B W and Ostlie D A 1996 *An Introduction to Modern Astrophysics* (Reading, MA: Addison-Wesley)
[2] Drury L O'C, Aharonian F A and Volk H J 1994 *Astron. Astrophys.* **287** 959
[3] Gaisser T K *Cosmic Rays and Particle Physics* (Cambridge: Cambridge University Press)
[4] Ginzburg V L and Syrovatskii S I 1964 *The Origin of Cosmic Rays* (New York: Pergamon)
[5] Hillas A M 1984 *Ann. Rev. Astron. Astrophys.* **22** 425
[6] Lagage G and Cesarsky C 1983 *Astron. Astrophys.* **118** 223
[7] Mewaldt R A *et al* 2001 *Space Sci. Rev.* **99** 27

Chapter 6

Gamma-ray observations of the Crab Nebula

6.1 Significance

The Crab Nebula, the SNR that resulted from a supernova explosion some 900 years ago, is one of the most important sources in high energy astrophysics. It has been said that half of all high energy astrophysics can be found in the Crab Nebula. As one of the few sources that has been observed for nearly a millennium, and at all wavelengths from longwave radio to VHE gamma rays, it is the best studied source in the cosmos. It was one of the first radio sources detected, it is one of the strongest x-ray sources, it was the first SNR to be clearly identified with a pulsar, it was one of the first gamma-ray sources detected (from balloon-borne telescopes), and it continues to provide glimpses of new astrophysical processes. For a time it was the fastest known radio pulsar (33 ms); at many wavelengths the radiation from the pulsar dominates the nebular emission. At optical wavelengths it is extraordinarily complex with many different phenomena superimposed (figure 6.1). It was the prototype source for synchrotron radiation by cosmic electrons [19] and is the prototype for Compton-synchrotron emission from cosmic sources [9]. Although generally seen as a strong and steady source, it is variable on time scales of days in the complex volume near the pulsar.

The Crab Nebula was first seen by Chinese, Japanese, and Korean astronomers (actually astrologers) when it exploded on 4 July, 1054 AD; it may also have been seen by Native Americans in the southwestern United States (see historical note: Crab pictograph). From the oriental records it is possible to deduce its brightness and light-curve in the first two years after its outburst. It could be seen in daylight in the first three weeks but now it cannot be seen at night with the naked eye. For 700 years it was not observed; then the invention of the optical telescope rendered it detectable again. Its striated appearance caused it to be named the Crab Nebula by Lord Rosse, the Irish astronomer, who observed it in the 19th century.

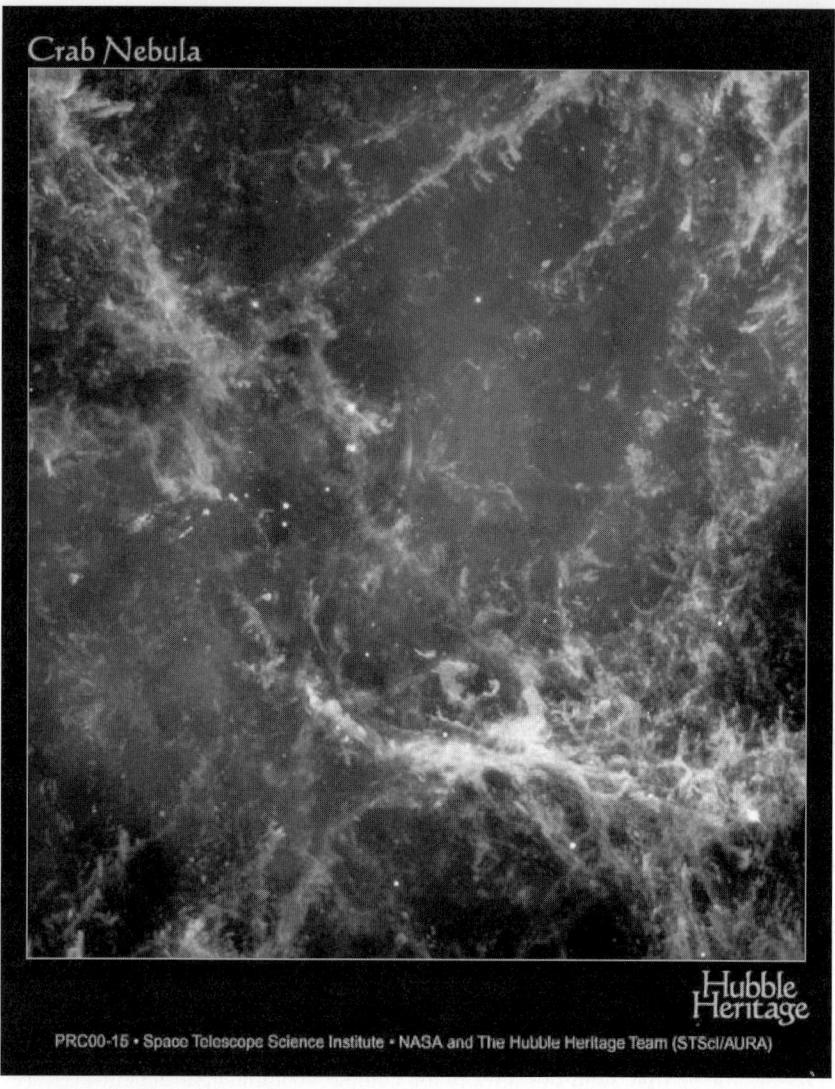

Figure 6.1. Optical image of the central part of the Crab Nebula as seen by the Hubble Space Telescope. The brightest image near the center is the pulsar. (Figure: STScI/NASA.)

6.2 Optical and x-ray observations

The apparent size of the Crab Nebula is a function of wavelength. The radio nebula shows the largest extent (4 arc-min diameter) whereas the optical nebula is only half this size. The x-ray nebula is smaller still. The emission at all of these wavelengths exhibits polarization and suggests the presence of synchrotron-radiating electrons. As the angular resolution of telescopes at all wavelengths has improved, the Crab Nebula has been shown to be increasingly complex.

Even before the discovery of the pulsar it had been realized that, in the region close to the two stars whose optical images are seen at the center of the optical image of the nebula, there was variable activity on a time scale of weeks. After the discovery of the pulsar these variations in the optical nebula were closely monitored. Faint optical wisps are seen to move outwards from the pulsar and there seems to be an underlying physical structure. The x-ray image from ROSAT and the images from the Hubble Space Telescope indicated that there was an axis of symmetry in the Crab Nebula which was oriented from the southeast to the northwest [10] and tilted at an angle of 20–30° to the sky plane. This implied that the Crab has a cylindrical structure whose axis is along its longest dimension.

The Hubble images suggested that there were jets emerging from the pulsar along the axis. The x-ray image indicated the presence of a torus at right angles to this axis. This picture was confirmed by Chandra in amazing detail (figure 6.2). It is now assumed that the cylindrical axis of the nebula is defined by the spin axis of the pulsar and that there is a rotating magnetic field close to the pulsar centered on this axis. The optical nebula is centered on the complex structure around the pulsar which must have a ambient magnetic field (from equipartition arguments) $\sim 3 \times 10^{-4}$ G. This is probably where the observed gamma radiation originates.

The most comprehensive model for the Crab Nebula is the magneto-hydrodynamic model of Kennel and Coroniti [15] which assumes spherical symmetry with a radial distribution of magnetic field strength. A critical factor in this model is the value of σ, the ratio of electromagnetic pressure to particle pressure at the radius of the pulsar wind shock.

6.3 Gamma-ray history

6.3.1 HE observations

The confused, but heroic, era of balloon-borne gamma-ray telescopes came to a conclusion in 1971 with the detection of a gamma-ray signal above 50 MeV from the direction of the Crab Nebula by the University of Southampton group [3]. The signal was only marginally statistically significant and then only when it was folded at the pulsar period. Hence, it was not immediately obvious that this was the decisive observation that would put the HE observation of discrete sources on a firm footing. However, the detection was to be confirmed in a number of balloon-borne experiments in the following years, including the Cornell University experiment using a gas Cherenkov telescope; this observation extended the measurements up to GeV energies.

The really definitive observation of the Crab Nebula, which established it as a HE source, came from the first satellite-borne spark chamber telescope, SAS-2, which clearly detected the source and measured its energy spectrum from 30 MeV to 500 MeV [18]. The pulsed component was fitted with a power law with spectral

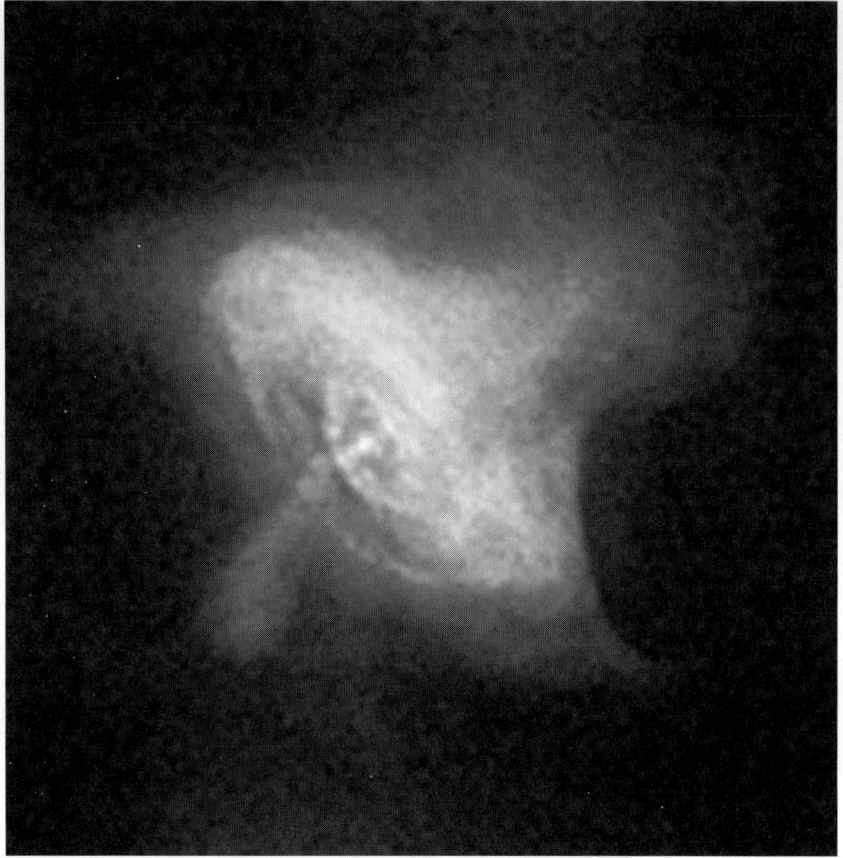

Figure 6.2. X-ray image of the region around the pulsar in the Crab Nebula as seen by Chandra. To avoid saturation the image of the pulsar itself has been blanked out. (Figure: Chandra/NASA.)

index -2.00 ($+0.60$, -0.55). Although there had been some suggestion from the balloon experiments of variability of amplitude in the overall signal, the SAS-2 results showed that the Crab pulsar was constant over its six month lifetime. The results were consistent with all of the emission coming from the pulsar, i.e. there was no nebular component. The positions of the two peaks in the pulsar light-curve were consistent with that measured at radio, optical, and x-ray frequencies. Two years later, the COS-B experiment achieved a much deeper exposure on the Crab source and identified a steady component, in addition to the pulsed component. Over the energy range from 50 to 500 MeV, the amplitude of the two components were approximately equal. It was not clear whether this unpulsed component came from the pulsar or the nebula. Up to 500 MeV, this steady

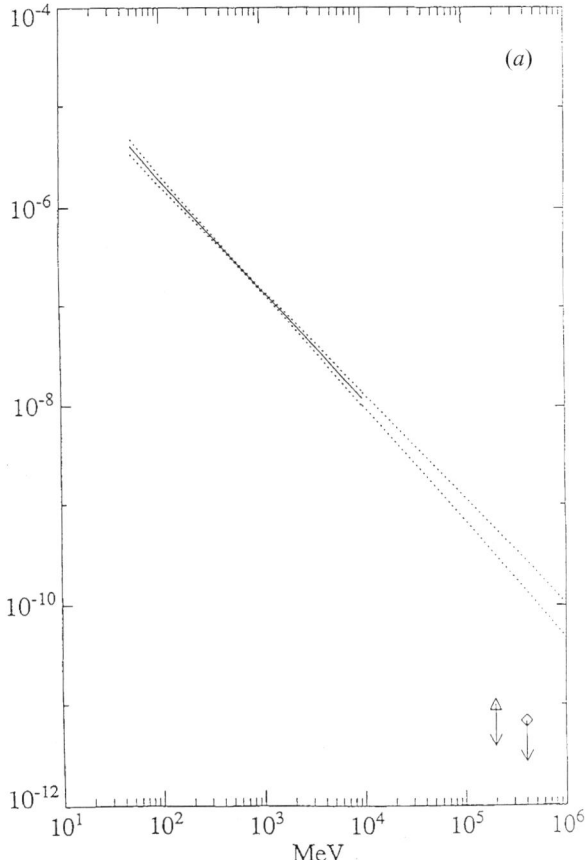

Figure 6.3. Early results on the Crab from EGRET and VHE observations. (*a*) The integral spectrum of the pulsed component of PSR 0531 + 21 (the upper limits from ground-based experiments are also shown). (*b*) The differential spectrum of the unpulsed emission. The full curve up to 10 GeV is from EGRET [18]; the full curve above 400 GeV is from Whipple [21]. The dotted lines are the $\pm 1\sigma$ uncertainties in each case. (Reproduced with permission from the *Astrophysical Journal*.)

component could be fitted by a power law of spectral index -2.7 ± 0.3 (compared to the pulsed component spectral index of -2.00 ± 0.10). Extrapolating these spectra to VHE energies indicated that it would be unlikely that the soft unpulsed signal would be detectable whereas the hard pulsar spectrum looked promising if the spectrum did not cut-off above 10 GeV (figure 6.3).

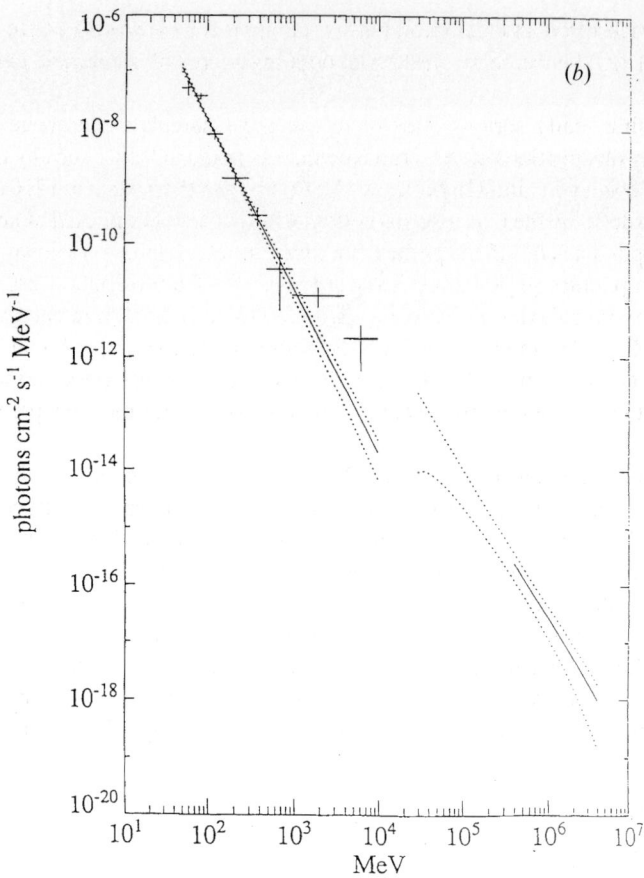

Figure 6.3. (Continued.)

6.3.2 VHE observations

Astronomy at TeV energies can be said to date from prescient predictions by Cocconi [8] of the detectability of TeV gamma rays from the Crab Nebula. Although his model for the gamma-ray emission was not correct (he overestimated the eventual detected flux by a factor of 1000), he sowed the seeds for the first serious atmospheric Cherenkov experiments to detect very high energy gamma rays from cosmic sources. His model of the Crab Nebula assumed that relativistic protons were produced in the initial explosion and were trapped in the nebula; collisions with the ambient gas produced charged pions, which subsequently decayed into relativistic electrons and these caused the observed synchrotron radiation. Gamma rays would come from the decay of the neutral pions produced in the same way. In practice, we now know that the lifetime against synchrotron and Compton loss is too great for electrons to have survived since 1054 AD; hence, there must be an ongoing source of acceleration within the nebula.

The first really serious attempt to use the Cherenkov technique to make gamma-ray observations of the Crab Nebula was made in the Crimea by the group from the Lebedev Institute in Moscow [7]; they began their work in 1960 and were inspired to do so by the Cocconi prediction. Their detector consisted of an array of 12 telescopes, each of 1.5 m aperture, mounted in sets of three on railway cars and directed in parallel (figure 1.4). They surveyed the Crab Nebula, Cassiopeia A, and some radio galaxies which were suspected to be sites of high energy particle activity. However, no excesses from the source directions were found and only upper limits were reported. The sensitivity achieved in the observations of the Crab Nebula was about two orders of magnitude below the flux predicted by Cocconi.

The development of a new model for the Crab Nebula was the catalyst for a new stage of development of atmospheric Cherenkov telescopes. If the amorphous radiation from radio to x-rays from the Nebula was due to synchrotron radiation by relativistic electrons, then these same electrons should Compton-scatter the photons, boosting them to gamma-ray energies [9]. The resultant gamma-ray spectrum would be most easily detectable at 100–1000 GeV energies, dipping sharply, thereafter, because the Klein–Nishina cross section (see appendix) comes into play (figure 6.4). The only free parameter in this model is the magnetic field which was assumed to be near the equipartition value.

6.4 Gamma source

6.4.1 The Crab resolved

Based on this prediction [9] of a hard gamma-ray component in the spectrum of the Crab Nebula, the Smithsonian Astrophysical Observatory built a 10 m optical reflector at the Whipple Observatory on Mount Hopkins in Arizona in 1968 for use as a VHE gamma-ray telescope; this was, for many years, the largest reflector built solely to do gamma-ray astronomy. As a first generation atmospheric Cherenkov detector, it was not successful in detecting a signal at more than 4σ from the Crab [6] or any other source.

The first indication that the Crab Nebula was to play the same pivotal role in VHE astronomy, as it had played at so many other wavelengths, came in the 1980s as a result of later observations at the Whipple Observatory. The 10 m reflector had been upgraded to act as an imaging detector and this resulted in the detection of a credible signal. A preliminary report of this detection was presented at the 19th International Cosmic Ray Conference in La Jolla, California in 1985 [4]. At a time when there was much confusion in the VHE detections of pulsars and binaries, this report drew little attention.

Selection of showers, based on the predicted properties of gamma-ray

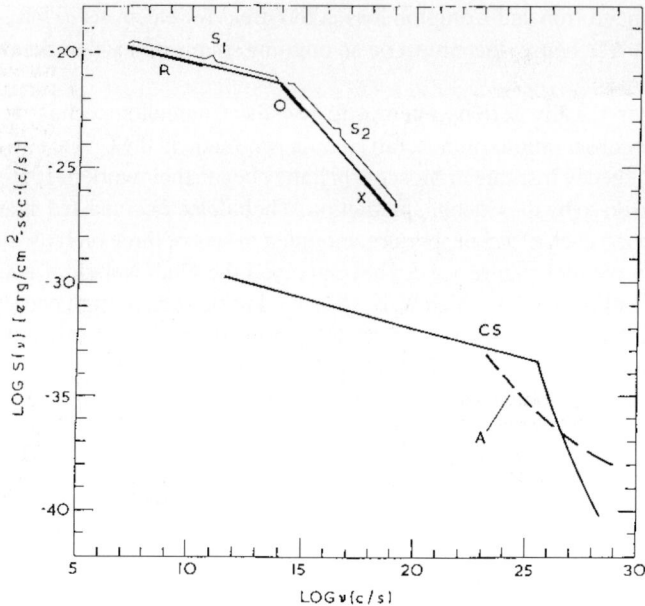

Figure 6.4. The Compton-synchrotron spectrum (CS) calculated by Gould [9] for the Crab Nebula; A is the spectrum expected from interaction with the 2.7 °K field [13].

showers, led to rejection of 97% of the background; a steady signal at the 9σ level was detected [22]. These observations on the Crab were extended in 1986–88 and resulted in a detection at the 20σ level which clearly made it statistically unassailable [21]. But there was no evidence for a signal from the pulsar, which satellite observations had suggested was the major gamma-ray source in the system. It should have been detectable at TeV energies based on the extrapolation of the pulsed spectra measured by SAS-2. It seemed unlikely that there could be an unpulsed component that was stronger than the pulsed component. It was initially assumed that the VHE signal, if real, must be pulsed and that there might be an error in the pulsar analysis.

The EGRET experiment on the CGRO [18], with its superior sensitivity to the previous space experiments, showed that the Crab gamma-ray source was more complex than previously supposed; the unpulsed MeV–GeV spectrum could not be fitted by a single power law. The resulting spectrum is shown in figure 6.5. The COMPTEL and EGRET data points, when combined with the VHE data points, can be fitted by the spectrum expected from a synchrotron-Compton model (see later) and match the extrapolation to lower energies from the VHE observations. This is an instance where a gamma-ray feature in a source was first seen in ground-based observations and was verified by a gamma-ray space experiment.

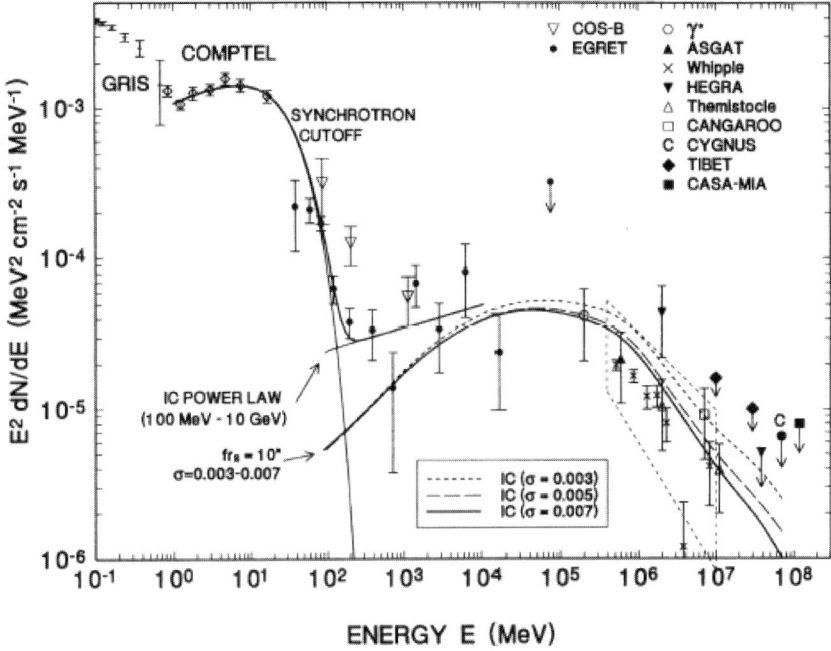

Figure 6.5. The unpulsed spectrum from the Crab showing measurements from GRIS, COMPTEL, COS-B, EGRET, and several ground-based telescopes. The data points from the space telescopes can be fitted with a synchrotron spectrum and the ground-based measurements with the resulting Compton spectrum for several values of σ [14]. (Figure: A Harding.) (Reproduced with permission from the *Astrophysical Journal*.)

The detection of an unpulsed signal, albeit weak, in EGRET observations finally convinced the skeptical space gamma-ray community that the unpulsed VHE signal was genuine and that ground-based observations were real. This strong detection of a weak steady VHE signal (about 0.2% of the observed cosmic ray background) was the first real evidence for the existence of TeV gamma-ray sources and opened up the new discipline of VHE gamma-ray astronomy.

6.4.2 The standard candle

From the COMPTEL and EGRET observations of the Crab, there is evidence for variability between 1 and 150 MeV [14]. However, at higher energies EGRET shows no evidence for variability. In the 300 GeV to 3 TeV range, the Crab Nebula, the archetypical plerion, is now considered a standard VHE candle (figure 6.6). It has been detected by eight ground-based gamma-ray telescopes using a variety of techniques. Of particular interest is the detection by the CANGAROO group; although their telescope is at latitude 31° south and the

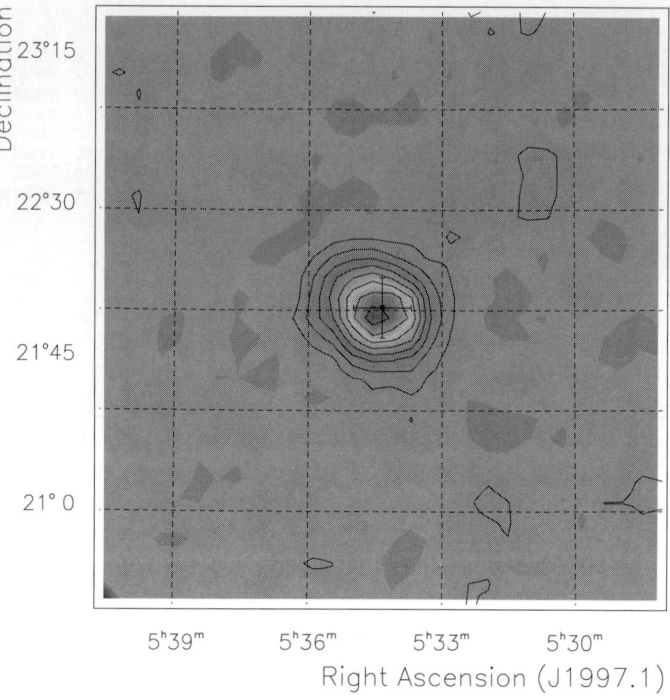

Figure 6.6. Isophoto contours of the probable location of the TeV source seen by the Whipple Observatory in the direction of the Crab Nebula. The source is not resolved but it is clearly coincident with the Crab (whose position is shown by a cross).

Crab Nebula is at declination $+21°$, they used their observations to establish the sensitivity of their telescope and to extend the energy coverage (by virtue of the observations at low elevations) [16]. The Crab Nebula was also seen by a conventional air shower array (the Tibet High Density Array at 4.5 km) where the energy threshold was 3 TeV. The energy spectrum agreed in shape with the Cherenkov measurements but was a factor of two to three times greater in absolute intensity. The Crab Nebula has also been detected by the Milagro experiment.

The Crab Nebula has been detected by four solar collector Cherenkov detectors, CELESTE, STACEE, Solar-2, and GRAAL, which have threshold energies as low as 50 GeV. There is no strong evidence of a pulsed signal even at these energies, an indication that the spectrum dips sharply at energies >10 GeV. The CELESTE result is compatible with an exponential cut-off with $E_0 = 26$ GeV [5]. The integral flux measured is $I(> 60 \text{ GeV}) = 6.2(+5.3, -2.3) \times 10^{-10}$ photons cm^{-2} s^{-1}. The measured spectra from several ground-based experiments are shown in figure 6.7.

Gamma source

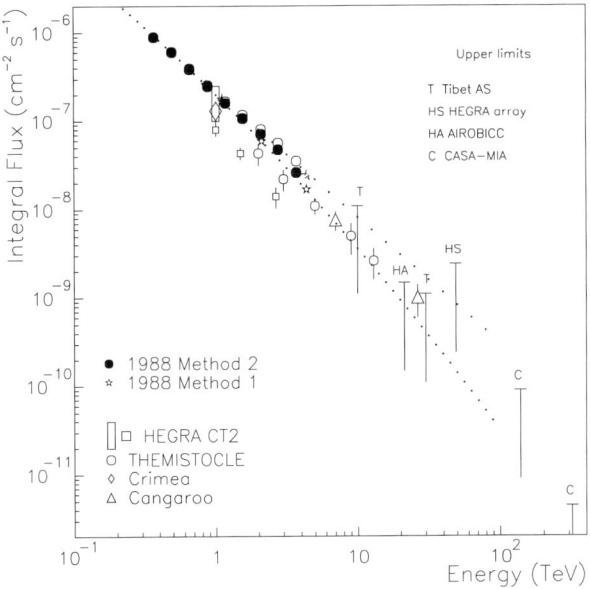

Figure 6.7. Spectra of Crab Nebula measured by several ground-based gamma-ray observatories [2].

Table 6.1. VHE gamma-ray flux from the Crab Nebula.

Group	Date	VHE spectrum (10^{-11} cm^{-2} s^{-1})	E_{th} (TeV)	Ref.
Whipple	1991	$(25(E/0.4 \text{ TeV}))^{-2.4 \pm 0.3}$	0.4	[21]
Whipple	1998	$(3.2 \pm 0.7)(E/\text{TeV})^{(-2.49 \pm 0.06_{stat} \pm 0.04_{syst})}$	0.3	[11]
HEGRA	1999	$(2.7 \pm 0.2 \pm 0.8)(E/\text{TeV})^{-2.60 \pm 0.05_{stat} \pm 0.05_{syst}}$	0.5	[1]
CAT	1999	$(2.7 \pm 0.17 \pm 0.40)(E/\text{TeV})^{-2.57 \pm 0.14_{stat} \pm 0.08_{syst}}$	0.25	[17]
CANGAROO	1998	$(2.01 \pm 0.36) \times 10^{-2}(E/7 \text{ TeV})^{-2.53 \pm 0.18}$	7	[20]
Tibet HD	1999	$(4.61 \pm 0.90) \times 10^{-1}(E/3 \text{ TeV})^{-2.62 \pm 0.17}$	3	[2]

There is remarkable agreement between the absolute fluxes and spectral shapes reported by several imaging ACTs; the results from the Whipple, HEGRA, CAT and CANGAROO experiments (as well as the Tibet Air Shower Array) are shown in table 6.1. There is no evidence for variability but the nature of the ground-based techniques is such that absolute fluxes are difficult to measure with confidence.

The EGRET observations span the synchrotron and Compton parts of the spectrum. The gamma rays up to 300 MeV are synchrotron radiation; unlike the rest of the spectrum they may exhibit some variability at this upper end of the spectrum corresponding to the highest energy electrons in the nebula. There is no evidence for any significant variation of this signal with time at TeV energies. The Compton-synchrotron model would not predict short-term variations although there might be a long-term secular decline.

The size of the nebula in gamma rays may also be a function of energy; with present angular resolutions this structure cannot be detected. Observations with the HEGRA telescope [12] determined that the apparent size of the gamma-ray nebula at a median energy of 2 TeV is <1.5 arc-min rms. For reference, the rms size of the radio nebula is 1.3 arc-min and the size predicted by the Compton-synchrotron models is <0.4 arc-min.

6.4.3 Interpretation

The spectrum of the Crab Nebula exhibits emission over a remarkably broad dynamic range, stretching from photons of energy less than 10^{-4} eV to those of nearly 10^{14} eV. The lower part of the photon energy spectrum (up to 100 MeV) can be explained as synchrotron emission from relativistic electrons within the nebula. The electron energies must be as high as 10^{15} eV. It is generally assumed that electron acceleration occurs in the termination shock of the pulsar wind; this occurs at a distance of 0.1 pc from the pulsar ($\sim$12 arc sec) [15]. From there, the electrons diffuse into the nebula. The presence of such high energy electrons in the luminous nebula inevitably leads to a Compton-scattered gamma-ray spectrum that extends to very high energies [9]. The low energy target photons can be either synchrotron photons, the 2.7 °K background or radiation from dust. In the VHE gamma-ray energy bands, the scattering is in the Klein–Nishina range with electrons having energy from 2–30 TeV and the soft photons having energies from 5×10^{-3} to 0.3 eV.

A number of calculations have been made of the Compton-synchrotron spectrum. The precise values depend on the distribution of the soft photons within the nebula, the distribution of the electrons, the magnetic field, and the distance to the nebula. The energy and spatial distribution is becoming increasingly well determined. A distance of 2 kpc is generally assumed. However, the only magnetic field estimate comes from the equipartition value (about 3×10^{-4} G). If the magnetic field is regarded as a variable, then the VHE gamma-ray measurements can be used to constrain it. In one model [11], the VHE gamma rays are scattered from the x-ray nebula close to the pulsar where the average magnetic field is 1.6×10^{-4} G and the electrons have energy >1 TeV (figure 6.8). Such electrons radiate synchrotron photons in the soft x-ray range so that an image of the nebula in soft x-rays also depicts the gamma-ray emitting region.

In another model [14], a detailed fit is made to the EGRET and VHE observations in terms of the magneto hydrodynamic model. The total spectrum cannot be fitted with a single value of σ which was assumed in [15] to have a

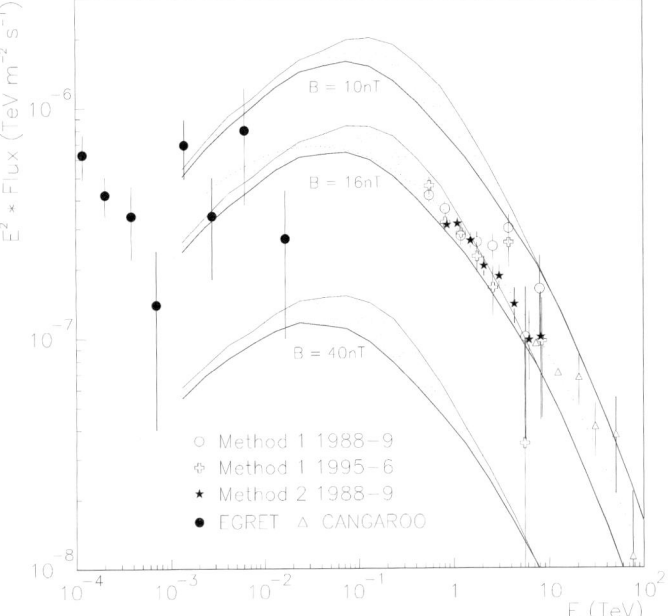

Figure 6.8. The VHE spectrum of the Crab Nebula as measured by EGRET [18] and the Whipple Observatory [11]. The various VHE data points are derived from two data sets using different methods of analysis to test for consistency. The predicted Compton spectrum for three values of magnetic field is also shown [11]. (Reproduced with permission from the *Astrophysical Journal*.)

value of 0.003. They find that they can fit the VHE spectrum very well but the observed flux is too low at sub-GeV energies. The best value of B in the region where the VHE gamma rays are radiated (the optical nebula) is 2.6×10^{-4} G and 1.3×10^{-4} G for the radio nebula. The variability observed in the 1–150 MeV range (the tail of the synchrotron spectrum) implies that there should be associated variability at energies >100 TeV. No measurements have been made at these energies yet.

The limits on the angular size of the TeV nebula [12] are still a factor of four above the size predicted by the canonical inverse-Compton synchrotron models and are barely compatible with the size predicted by alternative hadron models of production. In these, the energetic nucleons fill the radio nebula and produce gamma rays in nucleon interactions.

Figure 6.9. Pictograph from Chaco Canyon which appears to record the Crab Nebula explosion in 1054 AD.

Historical box: Crab pictograph

For the Anasazi people living in Chaco Canyon in the 10th and 11th centuries, the observation of the sun was of critical importance in determining the seasons. In this advanced society the duties of the Sun-priest demanded careful observation of the solstices. In 1054 AD therefore, it would be expected that he would be observing the sun rise in the weeks on either side of the summer solstice as he did every year. On the morning of 5 July on our calendar, just before dawn, on the northeastern horizon he would have seen the dramatic outburst that we call the Crab supernova explosion close to the crescent moon. So dramatic an event was this that he was moved to record it on a convenient overhanging rock face (figure 6.9). A celestial event of this magnitude surely merited a pictograph record. The star-like image, the crescent moon and the outline of a human hand can still be seen on the rock overhang. It is possible to think of this record as a scientific publication in which the event, the time, the position, the scale, and the brightness are all depicted. As in any good scientific publication, the author affixed his signature, in this case the imprint of his hand. Although fanciful, this conjecture illustrates the dramatic impact that a supernova must have had on an unsophisticated observer; with a better understanding of the nature of the phenomenon it should have an even greater impact.

References

[1] Aharonian F *et al* 2000 *Astrophys. J.* **539** 317
[2] Amenomori M *et al* 1999 *Proc. 26th ICRC (Salt Lake City)* **3** 456
[3] Browning R, Ramsden D and Wright P J 1971 *Nature* **232** 99
[4] Cawley M F *et al* 1985 *19th ICRC (La Jolla, CA)* **1** 131
[5] De Naurois M *et al* 2002 *Astrophys. J.* **566** 343
[6] Fazio G G *et al* 1973 *Astrophys. J. Lett.* **175** L117
[7] Chudakov A E *et al* 1965 *Transl. Cons. Bur., Lebedev Phys. Inst.* **26** 99
[8] Cocconi G 1959 *6th ICRC (Moscow)* **2** 309
[9] Gould R J 1965 *Phys. Rev. Lett.* **15** 577
[10] Hester J *et al* 1995 *Astrophys. J.* **448** 240
[11] Hillas A M *et al* 1998 *Astrophys. J.* **503** 744
[12] Hofmann W, Puhlhofer G and the HEGRA Collaboration 2001 *Proc. 27th ICRC (Hamburg)* vol 6, ed K H Kambert, G Heinzelmann and C Spiering (University of Hamburg) p 2403
[13] Jelley J V 1971 *The Crab Nebula* (Dordrecht: Reidel) p 32
[14] De Jager O C *et al* 1996 *Astrophys. J.* **457** 253
[15] Kennel C F and Coronoiti F V 1984 *Astrophys. J.* **283** 694
[16] Kifune T *et al* 1995 *Astrophys. J.* **438** L91
[17] Musquere A *et al* 1999 *Proc. 26th ICRC (Salt Lake City)* **3** 460
[18] Nolan P L *et al* 1993 *Astrophys. J.* **409** 697
[19] Shklovsky I S 1968 *Supernovae* (New York: Wiley–Interscience)
[20] Tanamori T *et al* 1998 *Astrophys. J. Lett.* **492** L33
[21] Vacanti G *et al* 1991 *Astrophys. J.* **377** 467
[22] Weekes T C *et al* 1989 *Astrophys. J.* **342** 379

Chapter 7

Gamma-ray observations of supernova remnants

7.1 Introduction

It is convenient to divide the discussion of gamma-ray observations of supernova remnants (SNRs) into two parts: plerions and shell-type SNRs. As we have seen in the previous chapter, there is no hard distinction between the two. VHE observations have been most pertinent to plerions and HE observations to shell-type SNRs. Neither set of observations have provided the smoking gun that would uniquely tie the origin of the cosmic radiation with SNRs.

7.2 Plerions

7.2.1 SNR/PSR1706-44

The radio pulsar, PSR1706-44, was discovered in 1992 and identified with the COS-B source 2CG 342-02; it has a period of 102 ms and a spin-down luminosity of 3×10^{36} erg s^{-1}. Gamma-ray pulsations were detected by EGRET at 100 MeV energies: it has a very flat spectrum at GeV energies. There is no evidence at these energies for unpulsed emission (see chapter 8). The pulsar is not detected at optical wavelengths although a weak pulsed x-ray source has been detected. There may also be a weak nebula (plerion).

The source was detected at TeV energies in 60 hr of observation with the CANGAROO telescope in the summer of 1992 [7]. This was the first season of operation of this high-resolution imaging telescope. Although weaker than the Crab, it showed promise to be the standard candle for TeV astronomy in the southern hemisphere. The energy spectrum was fitted with a power law with exponent -2.0. The flux above 1 TeV was $\sim 0.15 \times 10^{-11}$ cm^{-2} s^{-1}. No periodicity is seen at TeV energies nor was any other time variability noted; this is similar to the Crab Nebula. The observations are consistent with emission from an unresolved source of size <0.1°. The gamma-ray luminosity is 3×10^{33} erg s^{-1} which is approximately 10^{-3} of the spin-down energy of the pulsar and a factor of ten greater than the x-ray luminosity.

The TeV detection of this source was confirmed by the Durham group using a telescope at Narrabri, Australia. Observations with the CANGAROO-II telescope in 2000 extended the dynamic range of the TeV measurements and indicated a possible break in the energy spectrum near 1 TeV [8].

The best explanation for the unpulsed TeV emission from the PSR1706-44 source is in terms of an inverse-Compton synchrotron emission model assuming the existence of a plerion. If the x-rays (synchrotron) and the TeV gamma rays (inverse Compton) come from the same relativistic electrons, then the magnetic field in the plerion must be very weak, approximately 3×10^{-6} G, about the same value as the interstellar field. The observed spectrum would fit this model. However, the predicted intensity of the TeV emission is too low by a factor of ten and it is suggested that the TeV and x-ray emission may not come from the same physical region so that different electron populations are involved.

7.2.2 Vela

The CANGAROO group reported the detection of a 6σ signal from the vicinity of the Vela pulsar [14]. The integral gamma-ray flux above 2.5 TeV is 2.5×10^{-12} photons cm^{-2} s^{-1}. There is no evidence for periodicity and the flux limit is about a factor of ten less than the steady flux. The signal was originally thought to be offset (by 0.14°) from the pulsar position which made it more likely that the source was a plerion. However, this offset is no longer claimed by the CANGAROO group.

7.3 Shell-type SNRs

There is another group of shell-type SNRs which are observed at TeV energies; in these SNRs the progenitors are most likely electrons. These sources have not been detected at MeV–GeV energies.

7.3.1 SN1006

Perhaps because of its location in the southern hemisphere, the SNR associated with SN1006 has been somewhat neglected by modern observers [2] (see historical note: the supernova of 1006). In 1997, the CANGAROO collaboration detected TeV gamma-ray emission from this shell-type SNR [12]. The observations were motivated by the observation of non-thermal x-rays by the ASCA and ROSAT experiments and indicated a statistically significant excess from the northeast rim of the SNR shell. This detection of TeV gamma rays represented the first direct evidence of the acceleration of particles to TeV energies

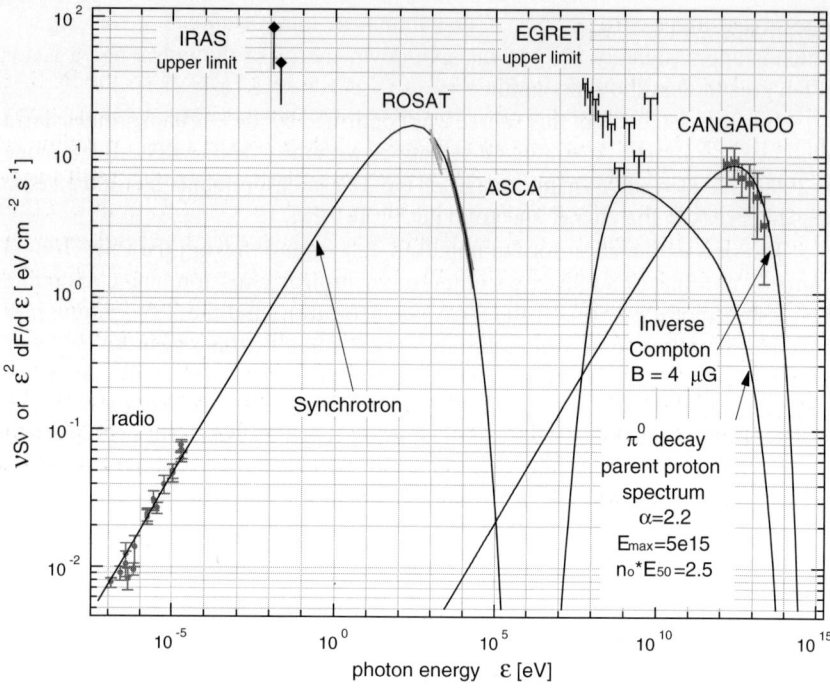

Figure 7.1. Observations of the north-east rim of SN1006 in the radio, soft, and hard x-ray bands as well as the CANGAROO measurements [13]. Also shown are upper limits from infrared and EGRET observations. The predicted gamma-ray spectrum assuming models of Compton-synchrotron and π^0 decay are shown as full curves. (Figure: T Tanimori.)

in the shocks of SNRs. However, the nature of the progenitor particles (electrons or protons) is not immediately obvious.

Later observations with the more sensitive CANGAROO-II telescope covered a wider range of energy (1.5–20 TeV) and permitted the derivation of a differential spectrum [13]. This had the following value:

$$F(E)dE = (1.1 \pm 0.4) \times 10^{-11}(E/1\text{ TeV})^{-2.3\pm 0.2} \text{ TeV}^{-1} \text{ cm}^{-1} \text{ s}^{-1}$$

where the errors are statistical only. A 30% systematic error should be included. A Compton model, where relativistic electrons scatter on the microwave photons of the 2.7 °K cosmic background, seems the best fit to the data. The observed TeV spectrum is consistent with a magnetic field of 4 μG (figure 7.1). This small value of magnetic field makes it difficult to produce the necessary electron acceleration and an alternative model suggests a field ten times this value. The matter density in the region of SN1006 is low (~ 0.4 cm^{-3}); hence, the predicted π^0 production is low and does not fit the observed gamma-ray spectrum.

7.3.2 RXJ1713.7-3946

This shell-type SNR was discovered in 1996 in an x-ray survey by the ROSAT satellite; its parameters are still somewhat uncertain. It has a characteristic dimension of 70 arc min, lies at a distance of 1–6 kpc and has an estimated age of 2000 to 10 000 years. Radio and infrared observations suggest it might be associated with a molecular cloud although it, like SN1006, is in a region of low interstellar density. TeV observations were motivated by the observation of a hard x-ray power-law spectrum by ASCA; these observations suggested it might be similar to SN1006 although its x-ray flux is three times brighter. The detection of TeV gamma rays from this relatively unknown SNR was reported by the CANGAROO group in 1999 [9]. The gamma-ray flux above 2 TeV is 3×10^{-12} photons cm^{-2} s^{-1}. If the upper value for the distance is taken, this is the strongest galactic VHE source in the sky in absolute terms. Subsequent observations with CANGAROO-II gave a deep detection (14.3σ) and permitted the derivation of an accurate spectrum [4]

$$dF_\gamma/dE = (1.63 \pm 0.15 \pm 0.32) \times 10^{-11} E^{-2.84 \pm 0.15 \pm 0.20} \text{ cm}^{-2} \text{ s}^{-1} \text{ TeV}^{-1}.$$

There is no evidence of a break or turn-over in the spectrum. As with SN1006, the emission comes from one part of the shell (the northwest rim).

It is difficult to fit the observed spectrum with a standard inverse Compton synchrotron model because of the absence of a spectral cut-off. Assuming the collision of cosmic ray protons accelerated in the source with interstellar gas, a good fit to the observed spectrum can be obtained for $E_{cr} = 10^{50}$ erg, $n = 100$ protons cm^{-3} and $d = 6$ kpc. This source could represent the first detection of a SNR in which the progenitor particles are hadrons and, hence, the first detection of a cosmic ray source.

However, the SNR may be associated with the unidentified EGRET source, 3EGJ1714-3857 [6]. If this identification is correct, then the EGRET spectrum clearly does not match the hadron production model (figure 7.2) [11]. If the identification is incorrect, then a strong upper limit to the flux at 1 GeV can be derived which is again in conflict with the model.

7.3.3 Cassiopeia A

Cassiopeia A (Cas A) is the strongest source in the radio sky and was one of the first targets of VHE observations. The source is a classical shell-type SNR of diameter 2.2 arc min which is effectively point-like to a gamma-ray telescope. Its distance is 3.4 kpc. Cas A is believed to be 300 years old and there is no active pulsar at its center; however, the x-ray image of neutron star at its center has recently been recorded (figure 6.1). It is appropriate that it should have been eventually detected as a TeV source (albeit only after a very long exposure by the HEGRA group [10]). As with SN1006 and RXJ1713.7-3946, these observations were motivated by observations of a hard x-ray power-law spectrum. However,

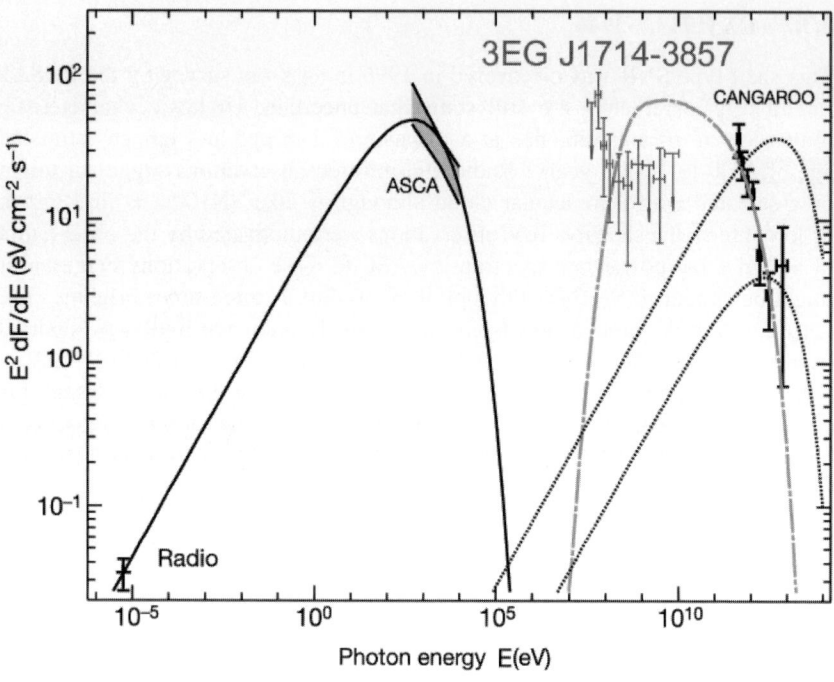

Figure 7.2. The multi-wavelength spectrum of RX J1713.7-3946. The full curve on the left shows the assumed synchrotron spectrum. The points measured by CANGAROO are shown as or the data points for 3EGJ1714-3857 [11]. The curves on the right are the model predictions for electron (short dashes) and hadron (large dashes) progenitors [4]. (Reprinted from Enomoto *et al* 2002 *Nature* **416** 823 with permission of Nature Publishing Group.)

for Cas A, the blast wave from the supernova outburst is seen to be expanding into a bubble region formed by previous outbursts from the central star which is believed to have been a red giant or Wolf–Rayet star.

The HEGRA observations were made over three years (1997–99) and comprised some 230 hr on the source. The flux above 1 TeV is given by

$$F_\gamma(> 1 \text{ TeV}) = (5.8 \pm 1.2_{\text{stat}} \pm 1.2_{\text{syst}}) \times 10^{-13} \text{ photons cm}^{-2} \text{ s}^{-1}.$$

The spectrum can be fitted with a differential spectral index of $-2.5 \pm 0.4_{\text{stat}} \pm 0.1_{\text{syst}}$. The uncertainty in the spectral index is sufficiently large that it is not possible to use it to distinguish between different models of gamma-ray production. The significance of the total detection was just less than 5σ and it is the weakest TeV source detected to date, i.e. approximately 3% of the Crab flux at these energies. The magnetic field is estimated as 10^{-3} G which is much larger than for SN1006 and implies a cut-off at a lower energy.

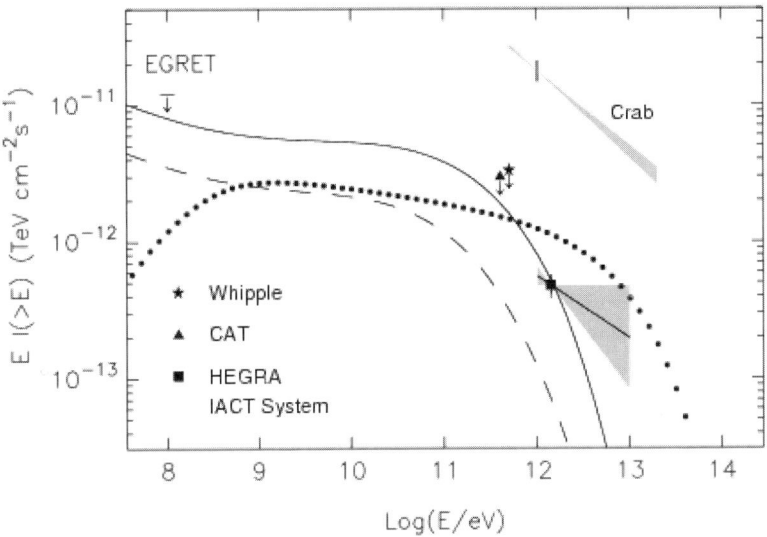

Figure 7.3. The gamma-ray observations of Cas A in the energy range from 100 MeV to 100 TeV. The HEGRA observations (1σ limits) are shown as shaded areas. The upper limits from EGRET, Whipple, and CAT are also shown. The predicted emission from Compton-synchrotron models are shown as full and dashed lines. The dotted curve (which can be re-normalized to fit the data) is a hadronic production model [10].

The interpretation of the observations of this important source is still controversial. The HEGRA detection must be reconciled with the non-detections at lower energies. Although it has been pointed out that the observed hard x-ray spectrum could be caused by bremsstrahlung by relativistic electrons, the general assumption is that the x-rays are synchrotron-radiated and that a Compton-synchrotron model is the best description of the TeV emission. This implies that the progenitors are electrons. An alternative model, which has greater uncertainties, predicts proton acceleration in the expanding shock wave with gamma-ray production by π^0 decay [10]. The observed spectrum (figure 7.3) can be better fit by this model. In practice, the gamma-ray spectrum may be the sum of a number of different processes and a full interpretation requires deeper detections over the full gamma-ray spectrum.

7.3.4 Other possible detections

Although they detected a number of discrete sources in the galactic plane (including pulsars), the SAS-2 and COS-B mission scientists did not produce any evidence for the emission of HE gamma rays from SNRs. Before launch it was hoped that EGRET would play a major role in the detection of HE gamma-

Table 7.1. Supernova remnants detected by EGRET.

SNR	Flux ($\times 10^{-8}$ cm^{-2}·s^{-1})	EGRET source	Distance (kpc)	Size (arc-min)
W28	56	2EG J1801−2312	1.8–4.0	42
W44	50	2EG J1857+0118	3	30
γ Cygni	126	2EG J2020+4026	1.8	60
IC443	50	2EG J0618+2234	0.7–2.0	45
Monoceros	23	2EG J0635+0521	0.8–1.6	220

ray emission from SNR and, hence, in establishing SNR as the prime source of hadronic cosmic rays. The observation of some extended sources, whose topology matched that seen at other wavelengths and whose energy spectra show the characteristic π^0 bump, would significantly advance the modelling of SNR. In fact, the observation of SNRs was one of the few disappointing results to come from the CGRO mission. The reasons are easy to understand; the relatively poor angular resolution of telescopes at 100 MeV energies, coupled with the strong background of gamma rays from the galactic plane, decreased the sensitivity to extended sources.

Although not nearly as well established as the association with radio pulsars, there is still a possible identification of several EGRET sources with well-known SNRs. Because such identifications could point to the SNRs as the source of cosmic ray acceleration, these claims have received much attention. The most prominent identifications [5] are listed in table 7.1, together with the source flux, $F_\gamma (> 100 \text{ MeV})$, and the approximate distance and angular size which are based on measurements at other wavelengths. The Third EGRET Catalog [6] lists seven other sources as tentative identifications with SNRs; these include Vela, Crab, CTA 1, S147, G312.4-0.4, Kes67, and G40.5-0.5. Of these, only the Crab is a pure plerion. As the study of these EGRET sources has been refined, particularly at higher energies, the identification with some of these sources, e.g. γ Cygni, has been called into question.

The brightest of these objects occur in dense regions that have been identified with molecular clouds. High densities of gas are required to explain the EGRET emission, of order 100 g cm^{-3}. It has been suggested that the MeV–GeV spectra of these sources may be more appropriate to that of radio-quiet pulsars so that the identification with shell SNR are somewhat in doubt.

There is no clear feature in the energy spectra (typically power laws with rather flat spectra) that suggest hadronic production. Subsequent work has shown that other processes, e.g. bremsstrahlung, inverse Compton, may provide better explanations and, hence, that hadronic production may not be required. If the acceleration of cosmic rays to energies of 100 TeV, and above, is to occur in these

Table 7.2. VHE observations of shell-type SNRs.

Object name	Observation time (min)	Energy (TeV)	Integral flux ($\times 10^{-11}$ cm^{-2} s^{-1})
Tycho	867.2	>0.3	<0.8
IC443	1076.7	>0.3	<2.1
	678.0	>0.5	<1.9
W44	360.1	>0.3	<3.0
W51	468.0	<0.3	<3.6
γ-Cygni	560.0	>0.3	<2.2
	2820.0	>0.5	<1.1
W63	140.0	>0.3	<6.4
SN1006	2040.0	>1.7	$0.46 \pm 0.6_{stat} \pm 1.4_{sys}$

sources, then they should also be strong sources of 1 TeV gamma rays. This is not the case.

VHE gamma-ray telescopes have much better angular resolution than EGRET, reducing the source confusion associated with any detection. Also, because the diffuse galactic gamma-ray emission has a relatively steep spectrum, proportional to $E^{-2.4}$–$E^{-2.7}$, compared with the expected approximate $E^{-2.1}$ spectrum of gamma rays from secondary pion decay, contamination from background gamma-ray emission should be less in the VHE range. Thus, searches for emission from shell-type SNRs have been a central part of the observation program of VHE telescopes.

Historical note: supernova of 1006

Most of the historic supernovae were recorded as dramatic optical events by early chroniclers of the skies in the northern hemisphere. It is remarkable that this object, whose outburst occurred during the Dark Ages, was noted at all since its declination was at $-41°$. There are many, although somewhat ambiguous, references to this outburst in the Arabic and far-Eastern records but the most remarkable record must be that made by the monks of St Gallen in Switzerland [2]. Because of the extremely low declination of the source for a northern hemisphere observer and the fact that the monastry is in a valley, it is possible to fix the position of the supernova relatively precisely and to remove any ambiguity in the identification with the modern position of radio supernova remnants.

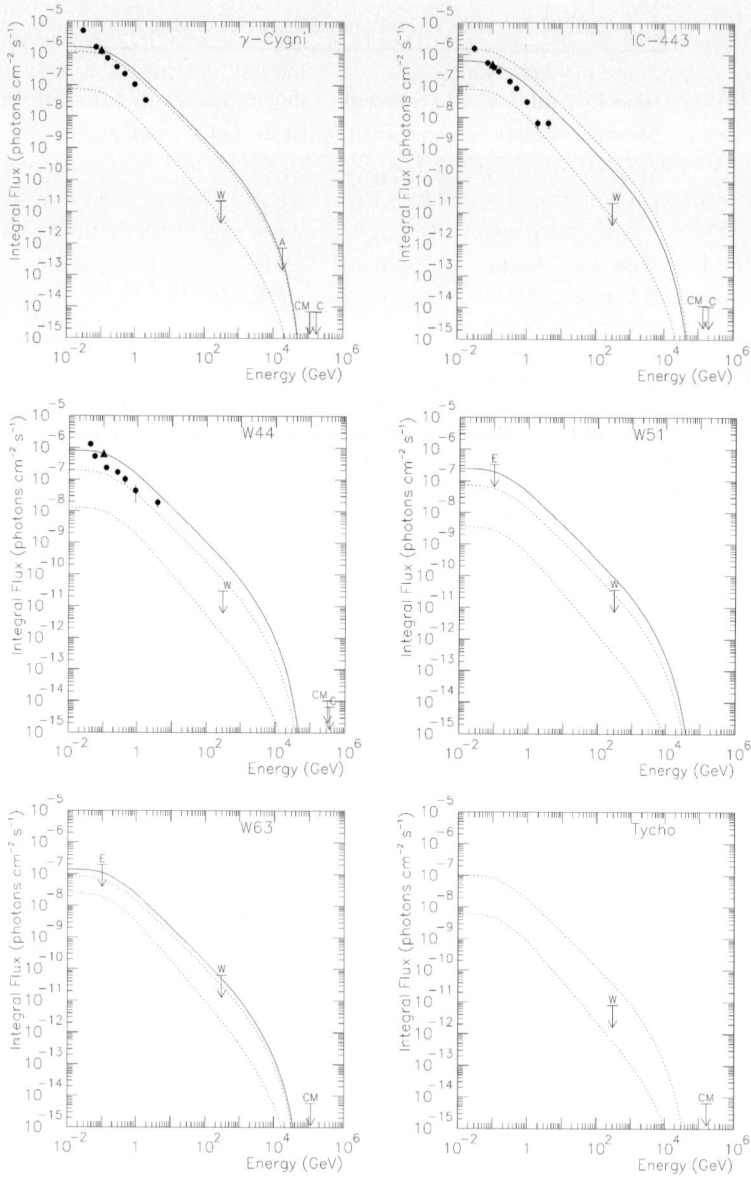

Figure 7.4. Integral spectra and upper limits from EGRET for several SNRs [5]. Also shown are upper limits from ground-based experiments: Whipple (W), CM (CASA-MIA), and Cygnus (C). The full curves are extrapolations from the EGRET integral data points [1]; the dotted lines are conservative estimates of the range of fluxes expected using the models of Drury *et al* [3].

The Whipple Observatory has reported the results of observations of six shell-type SNRs (IC443, γ-Cygni, W44, W51, W63, and Tycho) selected as strong gamma-ray candidates based on their radio properties, distance, small angular size, and possible association with a molecular cloud [1]. The small angular size was made a requirement due to the limited field of view (3° diameter) of the Whipple telescope. VHE telescopes can also detect fainter gamma-ray sources if they are more compact, because they can reject more of the cosmic ray background. IC443, γ-Cygni, and W44 are also associated with EGRET sources. Despite long observations, no significant excesses were observed, and stringent limits were derived on the VHE flux (see table 7.2 and figure 7.4).

References

[1] Buckley J H *et al* 1998 *Astron. Astrophys.* **329** 639
[2] Clark D H and Stephenson F R 1977 *The Historical Supernovae* (New York: Pergamon)
[3] Drury L O'C, Aharonian F A and Volk H J 1994 *Astron. Astrophys.* **287** 959
[4] Enomoto R *et al* 2002 *Nature* **416** 823
[5] Esposito J A *et al* 1996 *Astrophys. J.* **461** 820
[6] Hartman R C *et al* 1999 *Astrophys. J. Suppl.* **123** 79
[7] Kifune T *et al* 1995 *Astrophys. J.* **438** L91
[8] Kushida J *et al* 2001 *Proc. 27th ICRC (Hamburg)* **6** 2425
[9] Muraishi H *et al* 2000 *Astron. Astrophys.* **354** L57
[10] Puhlhofer G and the HEGRA Collaboration 2001 *Proc. 27th ICRC (Hamburg)* **6** 2451
[11] Reimer O and Pohl M 2002 *Astron. Astrophys.* **390** L43
[12] Tanimori T *et al* 1998 *Astrophys. J.* **497** L25
[13] Tanimori T *et al* 2001 *Proc. 27th ICRC (Hamburg)* **6** 2465
[14] Yoshikoshi T *et al* 1997 *Astrophys. J. Lett.* **487** L65

Chapter 8

Gamma-ray pulsars and binaries

8.1 General properties of pulsars

Although it is now more than 30 years since the discovery of the first radio pulsars, the topic is still an active one and there are more unanswered questions than there are satisfactory answers. Nearly 1500 isolated pulsars have been cataloged from radio searches; however, emission has been detected from only a small number of them over the entire band of the electromagnetic spectrum from radio wavelengths to high energy gamma rays.

Pulsars are divided into two categories: rotation-powered and accretion-powered. The former are generally detectable at radio wavelengths and the latter at x-ray wavelengths. The emission processes in the latter are generally thermal and of less interest to the high energy astrophysicist. Since they are always found in binaries, their discussion is deferred until later (section 8.5 below). There is another category known as millisecond pulsars which form an intermediate category.

It is generally agreed that rotation-powered pulsars are formed in supernova explosions when the cores of massive stars (>8 $M_\odot$) collapse [10]. The creation rate in the Galaxy is about one every 100 years. The current galactic population is, therefore, large. There is little question now as to the basic nature of pulsars, i.e. they are rotating neutron stars, their period, P, is the period of rotation, and the magnetic field of the neutron star, B, plays an essential role in producing the observed non-thermal radiation. The neutron star has a mass of approximately 1.4 $M_\odot$, a radius of 10 km and a magnetic field of 10^{12} G. The rotation energy

$$K = 1/2 I\omega^2 = 2\pi I/P^2$$

where I is the the moment of inertia. $I = 2/5 MR^2$ where $M =$ the mass and R is the radius. For $M = 1.4$ $M_\odot$ and $R = 10^6$ cm, $I = 1.1 \times 10^{45}$ g cm^2. The rate of rotation energy loss,

$$dK/dt = 4\pi^2 (I/P^3) \, dP/dt.$$

Table 8.1. Table of pulsar physical parameters.

Parameter	Units	Typical	Crab pulsar
Magnetic field	G	1×10^{-12}	4×10^{-12}
Rotation rate	rotations s^{-1}	1	30
Radius	cm	10^6	10^6
Mass	$M_\odot$	1.4	1.4
Moment of inertia	g cm^2	10^{45}	10^{45}
Electric field	V cm^{-1}	6×10^{12}	6×10^{10}

P can range from a millisecond to a few seconds. The rotation period of all pulsars is gradually increasing which is consistent with their loss of rotation energy. The characteristic age, τ, is given by

$$\tau = P/(2\mathrm{d}P/\mathrm{d}t).$$

For the Crab pulsar, $P = 0.0333$ s, $\mathrm{d}P/\mathrm{d}t = 4.21 \times 10^{-13}$ s s^{-1}, and $\mathrm{d}K/\mathrm{d}t = 5 \times 10^{38}$ erg s^{-1}. The radiation luminosity is assumed to be proportional to the rotation energy loss. Beyond these general properties, there is no general agreement about the nature of pulsars or their radiation mechanisms.

Some of the general physical parameters derived for a 'standard' pulsar are given in table 8.1, together with those derived for the Crab pulsar [11].

Gamma-ray studies at HE energies have added to the canon of knowledge about only a small number of pulsars but, as always, these observations are critical, as gamma-ray emission stretches the models to their limits. As yet, their contribution to the unsolved questions is still largely potential. After some marginal reports of detection in early experiments, only upper limits have been reported in the VHE band.

It is clear that the HE gamma-ray luminosity is high, in some cases more energy is given off at gamma-ray energies than in any other band and, hence, that gamma-ray observations may play a key role in unravelling the pulsar mystery. It is the relative weakness of current gamma-ray detection techniques, particularly in relation to radio techniques, that limits the gamma-ray catalog, not the inherent luminosity. Some measure of the weakness of the observed signals comes from the realization that the EGRET telescope only detected one photon every two hours from PSR B1055-52.

Table 8.2. EGRET-detected gamma-ray pulsar parameters.

Pulsar	Period (s)	Spin-down (10^{-15} s s^{-1})	Spectral index	Energy flux ($\times 10^{-10}$ erg cm^{-2} s^{-1})
Crab	0.033	421	2.15	10
B1951+32	0.040	5.85	1.74	2.4
B1706-44	0.102	93	1.72	8.3
Vela	0.089	125	1.70	71
B1055-52	0.197	5.83	1.18	4.2
Geminga	0.237	11.0	1.50	37

8.2 Gamma-ray observations

8.2.1 General characteristics

Most of what we know about the gamma-ray emission from pulsars has come from observations by the OSSE, COMPTEL, and EGRET instruments on the Compton Gamma Ray Observatory. Prior to the launch of the CGRO in 1991, the Crab and Vela pulsars were the only known sources of pulsed 100 MeV emission. Geminga was one of the strongest 100 Mev *unidentified* sources but its identity as a pulsar was only revealed during the EGRET mission (see historical note: Geminga in chapter 9). The COS-B source, 2CG342-02, was also known as a 100 MeV source, but it took the EGRET experiment to identify it with the pulsar, PSR B1706-44. Two other EGRET sources were identified with the pulsars, PSR B1055-52 and PSR B1951+32, on the basis of their positional coincidence and pulsed emission. PSR B1509-58 is technically a gamma-ray emitting pulsar but it is only seen at energies up to 1 MeV, hence it is not discussed here. There are tentative associations in the EGRET database with three other pulsars but the statistical significance is marginal (for each of these three, it is some three orders of magnitude less than that for the other seven). It is also suggested that some number of the unidentified EGRET sources may be pulsars whose radio beams are not pointing in our direction and whose signals are not strong enough for a periodic signal to be detectable in the EGRET database (see chapter 9). In retrospect, we know that the Geminga pulsar is so strong that it could have been found in a blind period search. In fact, it was only found after the period had been determined from x-ray measurements. It was subsequently also seen in the archival COS-B database.

The general characteristics of the seven gamma-ray pulsars are that they have flat spectra, that they are steady emitters over long time intervals, and that their gamma-ray luminosity is much less than the rotational energy loss. Some of the most important parameters of the six HE gamma-ray pulsars are summarized in table 8.2.

Gamma-ray pulsars are remarkable in that they give us the ability to measure certain parameters about the rotating neutron star, its environment, and its distance. We can thus hope to determine why it is that these pulsars are detectable in gamma rays and not others. For observability, the most important property is $\Omega_p = (dK/dt)/4\pi d^2$, where dK/dt is the rotation energy loss rate and d is the distance. This is basically the rotation energy loss rate scaled for distance. Five of the gamma-ray pulsars have values of Ω_p greater than 3×10^{-9} erg cm^{-2} s^{-1}; no other pulsars have Ω_p values this high (figure 8.1). The Ω_p of the sixth gamma-ray pulsar, PSR B1055-52, is high but is exceeded by a dozen others which are not observed in gamma rays. This might imply that this pulsar is a more efficient gamma-ray producer, has a preferred beam orientation, or that the distance measured is incorrect.

8.2.2 Spectral energy distribution

The spectral energy distributions of the gamma-ray emitting pulsars have several common features. All of them have power spectra that peak at gamma-ray energies and all have a definite cut-off energy. Hence, the peak luminosity is in gamma rays (figure 8.2). Although pulsars are a multi-wavelength phenomenon, when they are observed in gamma rays, the luminosity is orders of magnitude above that seen in most other wavebands. Had pulsars been first discovered at gamma-ray energies, the pulsar phenomenon might have been regarded as largely in the gamma-ray domain, with only auxiliary evidence coming from radio observations. Over a large part of the spectrum their emission is characterized by a power law. All of them turn over or break at some gamma-ray energy, usually in the GeV–TeV range.

There are clearly significant differences between the energy spectra of the six pulsars. The Crab pulsar shows a continuous spectrum from optical through gamma-ray wavelengths with a peak at 100 keV. There is a thermal component in the Vela pulsar spectrum and a sharp cutoff above a few GeV. PSR B1706-44 is characterized by two power laws with a turnover at 1 GeV. PSR B1951+32 has a flat spectrum at GeV energies but is not seen at VHE energies. Hence, it must have a very sharp cutoff above 10 GeV.

Geminga has very feeble radio emission, which has led to the suggestion that the gamma-ray beam is much wider than the radio beam, i.e. that it is only detected on the edge of the radio beam. There is a sharp turnover at a few GeV. It is also the closest known pulsar. There are two x-ray components, one of which is thermal.

The study of the high energy gamma rays elucidates not only the total power emitted from pulsars but also the acceleration of high energy particles by rotating neutron stars.

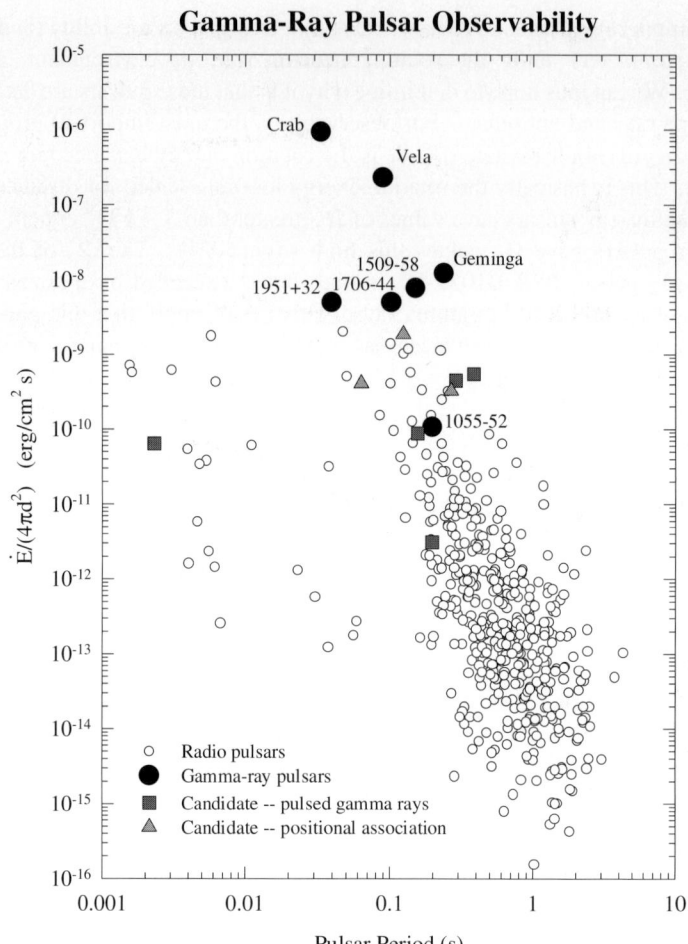

Figure 8.1. Pulsar gamma-ray observability (rotation energy loss scaled for distance) plotted as a function of period [15]. (Reprinted with permission from Thompson *et al* 1997 *AIP Conf. Proc.* **410**, *Proc. 4th Compton Symp. 1997* ed Dermer and Kurfess, pp 39–56.) (Figure: D J Thompson.)

8.2.3 Light curves

The light curves of the EGRET-detected pulsars are shown in figure 8.3, as seen at longer wavelengths and in the HE band. Only the Crab pulsar shows a light curve with the gamma-ray pulse in phase with the radio pulse. In contrast to the radio band, where single peaks are the norm, all six high energy gamma-ray pulsars have light curves consistent with double peaks separated by bridging emission. Geminga is the only pulsar where the separation is exactly 180°. The light curves

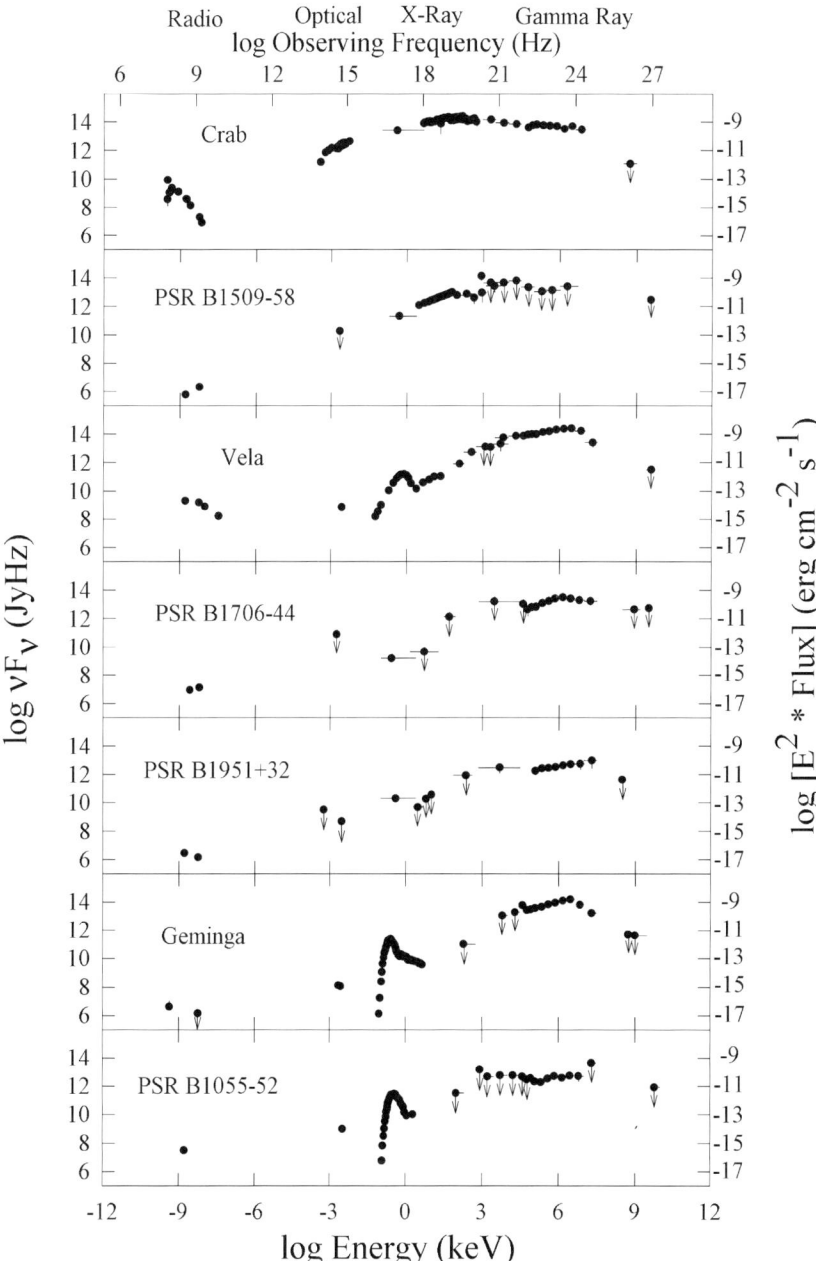

Figure 8.2. Power spectra for the known gamma-ray pulsars [15]. (Reprinted with permission from Thompson *et al* 1997 *AIP Conf. Proc. 410, Proc. 4th Compton Symp. 1997* ed Dermer and Kurfess, pp 39–56.) (Figure: D J Thompson.)

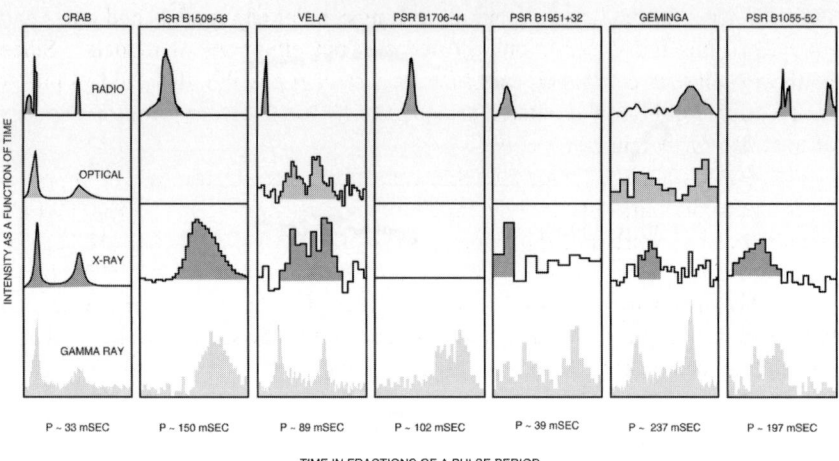

Figure 8.3. The light curves of the gamma-ray pulsars in wavebands from radio to HE gamma ray. (Figure: D J Thompson.)

of PSR B1055-52 and PSR B1706-44 are very broad and not very well defined. The early interpretation of the double pulse structure was that it arose from the observation of emission from the two opposite poles of the pulsar. It now appears that this is not the origin of the double-peak light curve but that it originates from a hollow cone of emission around a single pole.

For those pulsars with strong gamma-ray signals it is possible to measure the energy spectrum as a function of phase. There are marked differences which are not understood. For the Vela and Crab pulsars, the flattest spectra are measured in the bridging emission between the two peaks; for Geminga it comes at one of the peaks. At the highest EGRET energies (>5 GeV), the gamma-ray light curves change dramatically; for all of the pulsars for which there is sufficient statistics, the first of the peaks in the light curves disappears and the second pulse narrows significantly.

No pulsed emission is seen at VHE energies from any of the EGRET pulsars (figure 8.2). An early report of emission from the Crab pulsar has not been confirmed by more sensitive observations.

At HE energies the Crab pulsar is the only one of the six pulsars to have a detectable steady (unpulsed) component which could come from the surrounding supernova remnant. As discussed in the two previous chapters, at VHE energies only a steady signal is seen from the Crab, Vela, and PSR B1706-44 systems. This could be unpulsed radiation from the pulsar but is more likely from the associated plerion.

8.3 Models

Pulsar models fall under two general headings: polar cap models and outer gap models. Thus far, observations do not rule out either set of models. Since neither of these is completely satisfactory, it is also possible that the full pulsar explanation will be found outside these models. The geometry of these two sets of models is shown in figure 8.4(a).

The rotating strong magnetic field causes a strong electric field at the poles of the neutron star; this can be as much as 6×10^{10} V m^{-1} [5]. Particle acceleration is, therefore, inevitable. The velocity of the particles in the co-rotating magnetosphere will increase radially but cannot exceed the velocity of light. The magnetosphere only extends to the velocity of light circle beyond which there is a pulsar wind, containing both particles and magnetic field and connecting the pulsar to the surrounding medium.

8.3.1 Polar cap models

These are the oldest class of models dating back to the years immediately following the discovery of pulsars. In these models the particle acceleration and radiation occurs near the magnetic poles of the rotating neutron star [6]. There are a large number of variations in the models, in particular relating to the origin of the particles accelerated at the poles. In one set of models, the particles are pulled from the surface of the neutron star; in another, they are created by gamma rays producing electron–positron pairs. Acceleration of electrons to Lorentz factors of 10^7 is possible with subsequent emission of gamma rays by various mechanisms including curvature radiation and Compton scattering; these models predict cut-offs in the energy spectra above a few GeV which is consistent with the absence (thus far) of VHE pulsations. The polar cap models are successful in explaining the complex energy spectra of young pulsars but have more difficulty explaining the pulse shapes. Initially, these models assumed double pole emission to explain the light-curves but these were later modified to include emission from a hollow cone around a single pole (figure 8.4(b)). The same electrons that radiate the observed radio emission are predicted to produce the gamma rays; hence, for the polar cap models, one expects close correlation between the radio and gamma-ray properties.

8.3.2 Outer gap models

A vacuum gap can occur between the open field line and the null charge surface of the charge separated magnetosphere. These 'outer gaps' are the natural place for the acceleration of particles which then can radiate by curvature or inverse Compton scattering [4]. Again the models were initially bi-polar but there was little trouble in adapting the models to include double-peaked emission from a single pole. In general, outer gap models produce harder gamma-ray spectra but

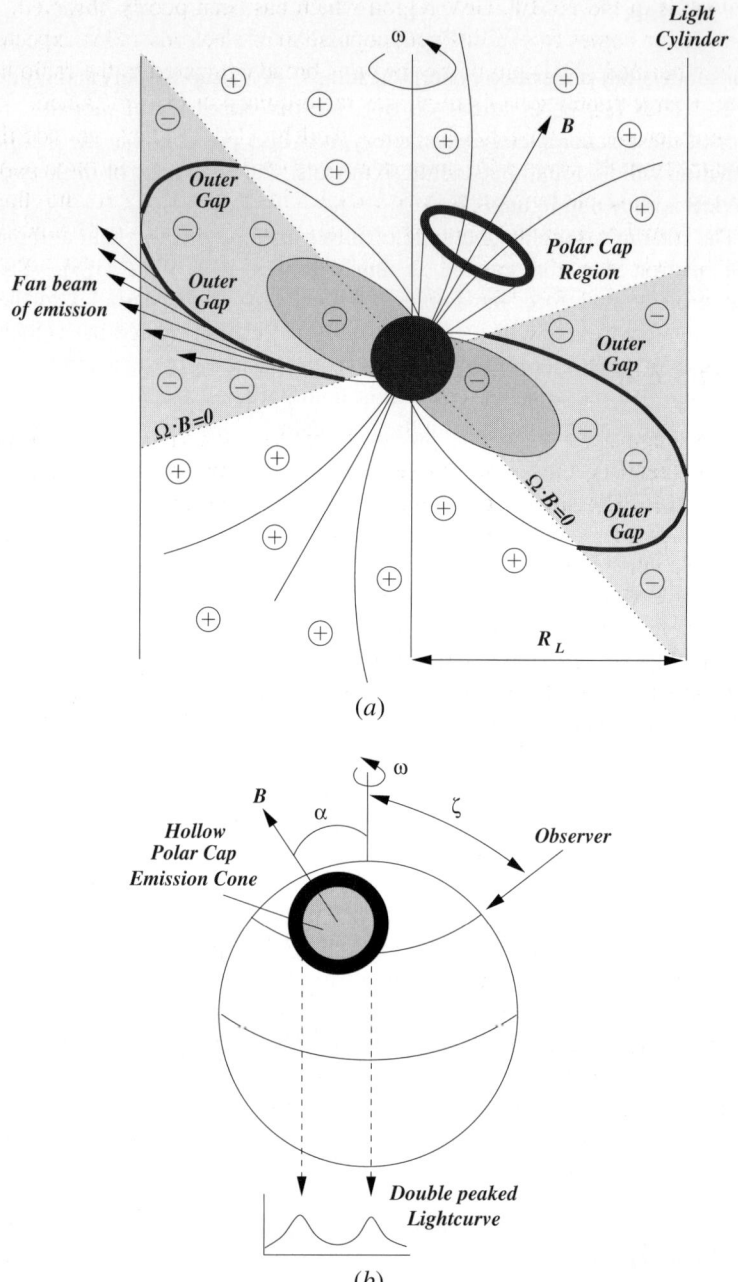

Figure 8.4. (*a*) Geometry of emission in polar cap and outer gap models; (*b*) double-pulse emission from hollow cone. (Figure: J Kildea.)

the cut-off is in the 10–100 GeV region which has been poorly observed. The radio emission comes from a different population of electrons and is expected to be tightly beamed. The gamma-ray beam is broad compared to the radio beam and thus a large population of gamma-ray radio-quiet pulsars is predicted.

An interesting variation on these models [14] predicts an additional component from the inverse Compton scattering of electrons on infrared photons in the gap. These photons could have energies in excess of a TeV and, thus, if detected, would be a unique signature of outer gap acceleration. However, the flux might only be 1% of that at GeV energies, so that detection must await the next generation of ground-based telescopes.

8.4 Outlook

Although gamma-ray studies of pulsars have provided vital new information for pulsar theorists, they have not yet permitted the two sets of models to be differentiated. The sample of gamma-ray emitting pulsars is so small that it is hard to identify generalities in their emission. Each new pulsar discovered has significantly added to our knowledge but has also skewed the overall interpretation of pulsar gamma-ray emission. With the small numbers involved, it is hard to determine what properties are significant. For instance, PSR B1509-58 has the highest pulsar magnetic field and the lowest energy cut-off, whereas PSR B1951+32 has the lowest field and the highest energy cutoff; is this correlation significant or a coincidence?

It is difficult to draw any definite conclusions as to the emission mechanism: in particular, it is not possible to differentiate between the favored polar cap or outer gap models. This may be possible with the next generation of gamma-ray telescopes. With its greater sensitivity compared to EGRET, GLAST will be capable of making phase-resolved observations of all the known gamma-ray pulsars. Also GLAST will detect some tens to hundreds of new pulsars so that definitive distinctions can be made in their populations. Some of these will be radio-quiet, like Geminga. In the brighter ones, it will be possible to determine the period. For others, it will be necessary to identify them with x-ray or optical sources. Combined observations by high energy space and ground-based instruments will also make definitive measurements of the energy spectra in the critical GeV–TeV region where cut-offs are expected (figure 8.5). Pulsars emit at a wide variety of wavelengths and it is expected that the final solution to the pulsar problem will require observations across the full spectrum.

8.5 Binaries

Half of the stars in the stellar population occur in some kind of binary association. The complex interaction between binary members, particularly when there is a compact object involved, was a key factor in the development of x-ray astronomy

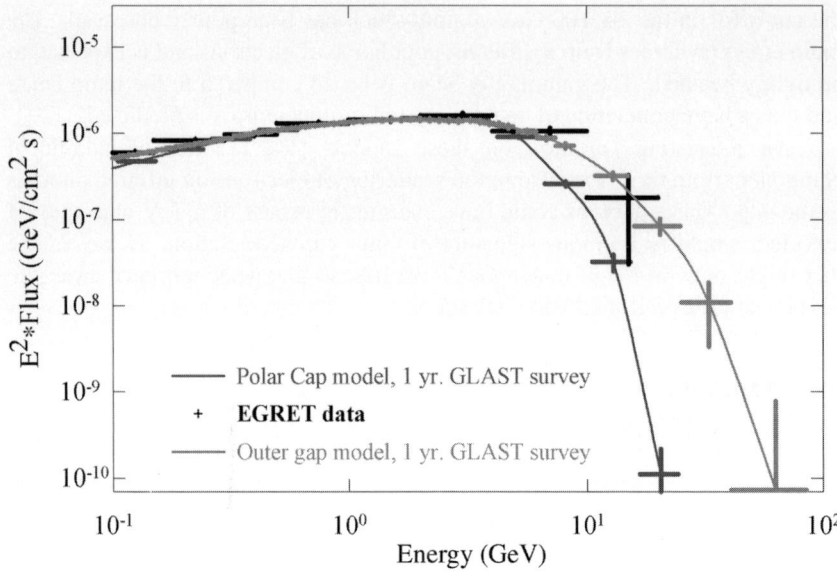

Figure 8.5. Models of the high energy gamma-ray spectrum of the Vela pulsar showing emission spectrum predicted by polar cap and outer gap models. The anticipated sensitivity of GLAST to these two predictions is also shown. (Figure: D J Thompson.)

since such sources were often the brightest x-ray sources in the Galaxy. The compact objects were white dwarfs, neutron stars, or black holes. Emission from these systems is highly time variable, with a variety of time constants from milliseconds to years. The variations can be in isolated flares or as quasi-periodic and periodic variations. These time signatures proved to be powerful tools for investigating these stellar systems.

The processes involved in the x-ray emission are associated with accretion and are largely thermal; there is not strong evidence for relativistic particle acceleration. Hence, it is not a great surprise that these sources have not proved to be important sources of HE gamma-ray emission. A search of the EGRET database for evidence of emission from known x-ray binaries gave no evidence that x-ray binaries might be HE sources [7]. At VHE energies, emission from binary sources has played an important historical role. At one time there appeared to be evidence that many binaries were sources of TeV gamma rays [2]. With the advent of new instruments with improved sensitivity, these early claims were not substantiated [19, 13].

There is one possible exception: a 5σ detection of a source has been reported by EGRET [17], which is positionally coincident with Centaurus X-3 (Cen X-3). This was detected during one observation of this region but not on others. There is weak evidence for emission at the 4.8 s period, assumed to be the spin of the pulsar, but no evidence for a correlation with the 2.09 day period, assumed to be the orbital period of the pulsar about its companion, an O-type supergiant. This is a luminous high mass x-ray binary ($L_x \approx 10^{38}$ erg s^{-1}) at a distance of 8 kpc. The gamma-ray source has a flat spectrum, with a differential spectral index equal to -1.81 ± 0.37, and with intensity less than $L_\gamma < 5 \times 10^{36}$ erg s^{-1}. The absence of a strong temporal correlation with the binary source makes the association with Cen-3 somewhat inconclusive; however, Cen X-3 lies close to the galactic plane and there is no strong AGN in the vicinity.

Historical note: Cygnus X-3

Historians of science will surely find the history of the gamma-ray detection of Cygnus X-3 a fruitful subject for the exploration of the sociology of scientific exploration as well as the development of statistical methods. The report of the detection of a gamma-ray signal at energies of 10^{14} eV by an air shower array in Kiel, Germany [16], which was quickly confirmed by the Havarah Park air shower array [9], led to an explosion in experimental activity at all gamma-ray energies. Because of the large luminosity implied, it was widely believed that Cygnus X-3 might be an important source of cosmic radiation.

In fact, Cygnus X-3 had already been reported as a HE gamma-ray source, based on SAS-2 observations [8], and as a VHE gamma-ray source, whose emission was correlated with the large radio outbursts [18]. There followed a spate of new reports of gamma-ray emission at energies ranging from 400 GeV to 10^{18} eV. Almost all the detections reported the detection of the 4.8 hr period. A pulsar period of 12.57 ms was reported from one VHE experiment; another experiment reported a period of 9.7 ms.

Interest grew when it was reported that the Kiel signal did not have the low muon content expected of electromagnetic air showers. However, skepticism in the reality of the gamma-ray detections also increased as it appeared that the detected signals of Cygnus X-3 were always of marginal significance (over ten decades of energy), the time variations were different in each experiment, and significance did not increase with exposure [1, 19]. New observations with improved telescopes produced only upper limits. There was no evidence for a signal in either the COS-B or EGRET experiments [12]. As the atmospheric Cherenkov technique was developed with improved flux sensitivity, the signal strength seemed to diminish. Large particle array experiments were built with factors of ten to a hundred improvement in flux sensitivity on the Kiel and Haverah Park experiments. No signal was detected [13].

With hindsight, it would be easy to dismiss the whole Cygnus X-3 phenomenon as a scientific red herring. It was not impossible that the source was bright but gradually faded. Sadly, in cosmic ray studies this is not an unfamiliar phenomenon where apparent sources have often disappeared as techniques have improved.

Cen X-3 had been one of the early binaries reported by VHE observations, but the reports depended on periodicities and seemed inconsistent [19]. The x-ray periodicities were not well determined, so that there were many degrees of freedom available for periodic analysis and these were not always fully accounted for. With the launch of CGRO in 1991, BATSE provided routine monitoring of binary sources and removed any ambiguity as to period. Later VHE observations by the Durham group [3], using these contemporaneous BATSE observations, did not see any evidence for periodicities but did confirm the existence of weak, but persistent, emission at energies >400 GeV at the 4.7σ level. Unfortunately, this does not agree too well with the EGRET observations, which saw emission at $E > 100$ MeV only as a flare lasting for two weeks.

Other binaries, which were reported to have anomalous VHE emissions, included Hercules X-1, 4U0115+63, Vela X-1, LMC X-4, and Scorpius X-1 [2].

Interest in HE and VHE emission from x-ray binaries had initially been triggered by observations of Cygnus X-3 (Cyg X-3) in the 1970s and 1980s (see historical note: Cygnus X-3). Cygnus X-3 is a well-studied and bright x-ray binary source with a well-established 4.8 hr period. Its outstanding characteristic is the emission of bright radio flares which last for days but occur at irregular intervals on a time scale of years. Cygnus X-3 is optically obscured by dust in the galactic plane and, hence, its study has been difficult. It is now believed that Cygnus X-3 is a microquasar, whose jet is directed towards the Solar System. Its status as a gamma-ray source is still undetermined and it remains an object of much mystery.

References

[1] Bonnet-Bidaud J M and Chardin G 1988 *Phys. Rep.* **170** 325
[2] Chadwick P M, McComb T J L and Turver K E 1990 *J. Phys. G: Nucl. Part. Phys.* **16** 1773
[3] Chadwick P M *et al* 2000 *Astron. Astrophys.* **364** 165
[4] Cheng K S, Ho C and Ruderman M A 1986 *Astrophys. J.* **300** 500
[5] Goldreich P and Julian W H 1969 *Astrophys. J.* **245** 267
[6] Harding A 1981 *Astrophys. J.* **245** 267
[7] Jones B B *et al* 1997 *Proc. 4th Compton Symposium (Willamsburg) (AIP Conf. Proc. 410)* ed C D Dermer, M S Strickman and J D Kurfess (New York: AIP) p 783
[8] Lamb R C *et al* 1977 *Astrophys. J.* **184** 271
[9] Lloyd-Evans J *et al* 1983 *Nature* **305** 784
[10] Lyne A G and Graham-Smith F 1990 *Pulsar Astronomy* (Cambridge: Cambridge University Press)
[11] Michel C 1991 *Theory of Neutron Star Magnetospheres* (Chicago, IL: University of Chicago Press)
[12] Mori M *et al* 1997 *Astrophys. J.* **476** 842
[13] Ong R 1998 *Phys. Rep.* **305** 94
[14] Romani R W 1996 *Astrophys. J.* **470** 469
[15] Thompson D J *et al* 1997 *Proc. 4th Compton Symposium (Williamsburg, VA) (AIP*

Conf. Proc. 410) ed C D Dermer, M S Strickman and J D Kurfess (New York: AIP) p 39
[16] Samorski M and Stamm W 1983 *Astrophys. J.* **268** L17
[17] Vestrand W *et al* 1997 *Astrophys. J. Lett.* **483** L49
[18] Vladimirsky B M, Stepanian A A and Fomin V P 1973 *Proc. 13th ICRC (Denver)* **1** 456
[19] Weekes T C 1993 *Space Sci. Rev.* **59** 315

Chapter 9

Unidentified sources

9.1 HE observations

In the very early days of balloon spark chamber gamma-ray telescopes, there were a plethora of claims for the detection of unidentified discrete sources. All of these were of marginal statistical significance. Only one was subsequently confirmed in satellite experiments and this, the Crab, was clearly a known source.

The short-lived SAS-2 mission detected three discrete sources: the Crab, Vela, and one which was unidentified until 20 years later (see historical note: Geminga). Identifications with an SNR and with an OB association (a dense concentration of young massive O and B stars) were suggested but not confirmed. The COS-B mission concentrated its observations on the galactic plane and in its final catalog recorded the positions of 30 discrete sources. Almost half of these were subsequently found to be due to incorrect subtractions of the diffuse galactic plane from the signal. The identified sources included Vela, the Crab, and 3C273. This still left the majority of the sources detected unidentified.

The Third EGRET Catalog of discrete sources [5] listed 271 sources. Point-source locations and fluxes were based on the first four years of observation of the CGRO mission and were determined using the maximum likelihood method. For a source to be included it had to have a statistical significance of more than 4σ; for sources that lay within $10°$ of the galactic plane, this level was increased to 5σ. This latter level was necessary because sources at these galactic latitudes were seen against the model of the diffuse galactic emission whose subtraction increased the uncertainty of the detection. In practice, since exposures across the sky were not equal and tended to be concentrated towards the galactic equator anyway, the probability of source detection was about equal across the sky. However, there were several instrumental and observational peculiarities in the EGRET mission which prevented complete uniformity in sensitivity from being achieved [10]. In fact, more detailed analyses of the EGRET database has led to a revision downwards in the number of sources in the Third EGRET Catalog. In particular, seven new 'sources' close to the strong, well-established, sources, Vela

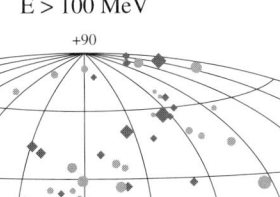

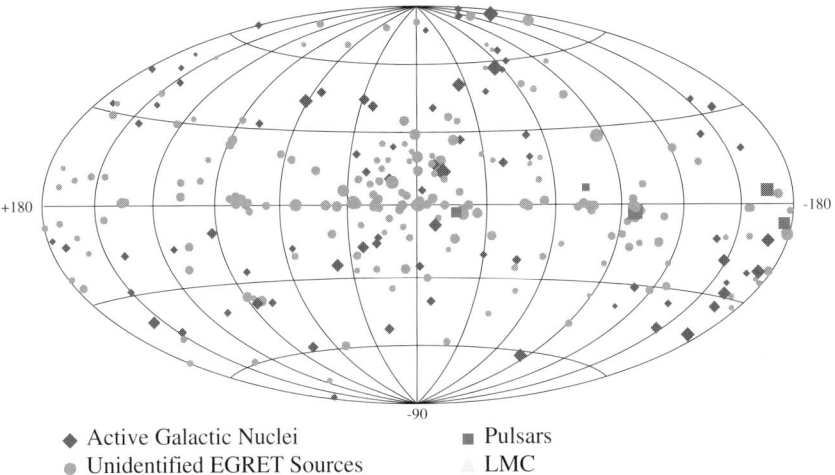

Figure 9.1. The Third EGRET Catalog of discrete sources [5]. The pulsars are shown as filled squares, the active galactic nuclei as filled triangles and the unidentified sources as filled circles. (Reproduced with permission from the *Astrophysical Journal*.)

and Crab, must be deleted. The angular resolution of the EGRET instrument is such that for the majority of the sources (just above the threshold of detectability) the location error box is of order 1.0° radius. The angular resolution of the EGRET telescope goes as $E_\gamma^{-0.5}$ so there is definite advantage in making maps at the highest energies where the signal is statistically significant. The GeV EGRET Catalog [7] lists more than 50 sources of gamma rays with energies greater than 1 GeV. Many of these sources are coincident with the 100 MeV sources but the positions are better determined.

9.2 Population studies

There is a clear concentration of sources towards the galactic plane in the map of discrete sources taken from the Third EGRET Catalog (figure 9.1). A similar concentration is seen in maps made in sky surveys in the visible, infrared, and x-ray wavelength intervals. In fact, 74 of the 271 sources in the Third EGRET Catalog lie within ±10° of the plane. It is generally thought that more than half of the sources in the catalog are galactic. However, only a small proportion of them (<5%) have been positively identified with known galactic objects. The nature of the majority of the objects is completely unknown and is one of the major mysteries, and unsolved legacies, of the EGRET mission.

Since more than half of the sources found by the COS-B mission were later found to be high points in the galactic diffuse emission, the EGRET discrete sources must be treated with some caution. The error boxes are large and there are many possible objects within them. The problem of identification with known objects at other wavelengths is compounded by the density of objects in the galactic plane, the uneven nature of the diffuse galactic plane distribution which must be subtracted, the possibility of source confusion or overlapping sources, the possibility that some sources are extended, the change of sensitivity of EGRET over the lifetime of the mission, the observed time variability of many of the sources, and the lack of independent verification of the detections by another gamma-ray telescope.

Attempts at identification follow two general lines: (1) statistical association of the distribution of a sub-class of sources with known galactic objects; or (2) positional and/or temporal association of an individual source with an object that is prominent or unusual at other wavelengths. The quest for individual identifications has become a major industry and is discussed in the next section.

A characteristic feature of extragalactic sources (although perhaps not unique to them) is that they are variable on time scales of weeks to years. Hence, an easy sub-division of the 170 unidentified sources is into those that are time variable and those that are not. The 120 sources in this latter category are plotted in figure 9.2 [2]. A further sub-division into those that are brighter or fainter than 2.4×10^{-7} photons cm^{-2} s^{-1} shows that the former are concentrated along the plane whilst the latter have a wider distribution but tend to cluster above the galactic center. The low-latitude sources have a hard differential spectral index on average of -2.18 ± 0.04, whilst the mid-latitude sources have indices of -2.49 ± 0.04. These mid-latitude sources are now believed to constitute a distinct population.

As seen from figure 9.2, the mid-latitude sources are almost coincident with a concentration of stars known as Gould's Belt. This is a large, relatively nearby, expanding disk of stars (about 1000 light-years in radius) that is inclined to the plane; the sun is about half-way to its outside rim. This could explain the asymmetric distribution of gamma-ray sources which mostly lie to the north of the galactic plane. The origin of Gould's Belt is unknown but it contains a large population of molecular clouds and supernova remnants; it also contains many young stars and that fact, and the ongoing expansion of the ring, suggests a violent event in the past at its center. The distribution of sources is consistent with typical distances of 100–400 pc, which suggests that they are relatively local and weak in absolute terms. By contrast, the sources that lie along the galactic plane must be on average 1–4 kpc from the Solar System.

A study of the low-latitude sources shows a general positional correlation with supernova remnants and O stars. However, individual source identifications have considerable uncertainty. Since the identified sources that lie at low latitudes are all pulsars, it is strongly suggested that the unidentified sources are radio-quiet pulsars (like Geminga); in these sources the radio beam may be narrower than the gamma-ray beam.

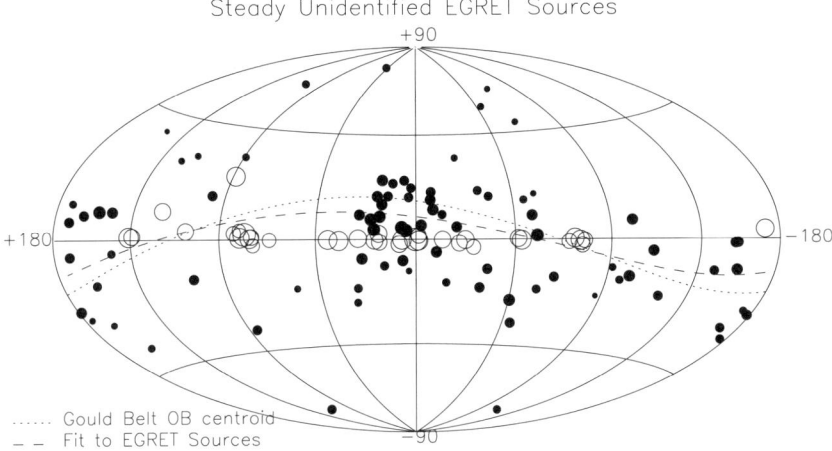

Figure 9.2. The steady unidentified sources detected by EGRET. The size of the points is proportional to the flux level of the source. The dotted line represents the center of the Gould Belt. The broken curve is the great circle that best fits the EGRET sources [2]. (Figure: N Gehrels.) (Reprinted from Gehrels *et al* 2000 *Nature* **404** 63 with permission of Nature Publishing Group.)

Table 9.1. Source populations.

Location	Low latitude	Mid-latitude
Distribution	Along plane	Concentrated north of plane
Intensity	Bright	Faint
Spectral index	-2.18 ± 0.04	-2.49 ± 0.04
Distance	1–4 kpc	100–400 pc
Power (erg s^{-1})	0.6–4.10^{35}	0.7–1.41×0^{33}
Position	Plane	Gould Belt
Sources	Radio-quiet pulsars	SNRs, O stars

The properties of the two populations of sources which are assumed to be galactic are listed in table 9.1. A third population of galactic sources has also been postulated with a broad distribution consistent with being in the galactic halo [3]; these would include some of the variable sources and would, therefore, be a somewhat different kind of object (perhaps 35 $M_\odot$ black holes accreting from the interstellar medium).

9.3 Individual identifications

Since there is a lull in HE gamma-ray observations in the interval between EGRET and GLAST, there has been considerable activity by astronomers at all wavelengths to identify the EGRET sources. Some of these are quite convincing with strong cases being made for identifications with individual pulsars and AGN. The scientific literature (particularly the proceedings of conferences) contains a number of claims for such identifications but many of them must be regarded as speculative. In some cases, there is an underlying time variation (periodic or flare-like) in both the observed gamma-ray signal and the radio, optical or x-ray signal in which case the identification can be regarded with some confidence. However, even here some caution must be exercised as periodic signals are notorious for their statistical shades of grey and the early literature is sprinkled with apparent detections which have never been confirmed. This is particularly so in the case of periodic sources such as x-ray binaries where the contemporary empherides are not well established so that extra (and often ill-defined) degrees of freedom are available in the data analysis. The most unambiguous identifications are with radio pulsars where the ephemeris is well established and the gamma-ray signal is strong.

Individual identifications are most interesting because they permit detailed modelling of the source and its emission processes. Identifications with pulsars (chapter 8), supernova remnants (chapter 7) and a binary x-ray source (chapter 8) are discussed in the appropriate chapters. Here we list just a few interesting cases; there are many more [1].

9.3.1 CG135+01

One of the strongest sources seen by COS-B and located close to the galactic plane, 2CG 135+01, still lacks a definite identification. Initially, there were two strong candidates within the COS-B error circle: the quasar, 4U0241+61; and the variable non-thermal radio source, GT 0236+610. However, the strong detection by EGRET (3EG J0241+6103) permitted a better position to be determined and this eliminated the quasar as a possible identification. The flux above 100 MeV, averaged over four years, was $F_\gamma = 9.2 \pm 0.6 \times 10^{-7}$ photons cm^{-2} s^{-1}; there was only weak evidence for variability. The spectrum was hard and could be fitted with a power law with spectral index -2.05 ± 0.06 [8]. It must turn-over above 30 GeV as it is not detected in VHE experiments.

The possible association of the gamma-ray source with the variable radio source is of great interest since, if real, it would imply a whole new class of gamma-ray sources. The radio source is identified with a massive B star (LSI+61°303) which may be one member of a binary. The radio signal is clearly variable with flares every 26.5 days and a possible four year modulation. Models that have been proposed included a binary system containing a pulsar with an eccentric orbit and a supercritical accretion model in which the emission comes at the periastron. There is not a clear correlation on these time scales with the gamma-ray signal which casts some doubt on the latter model. A variability correlation with the longer wavelengths would confirm the identification.

9.3.2 3EG J0634+0521: binary pulsar?

It has been suggested that 3EG J0634+0521 can be identified with the young binary pulsar, SAX J0635+0533, a relatively bright x-ray source with a hard spectrum whose optical counterpart is a Be star [6]. The orbital period is 11.2 days and the pulsar period is 34 ms. The spin-down power is $dE/dt = 5 \times 10^{38}$ erg s^{-1} which is remarkably similar to that of the Crab pulsar. As yet, the gamma-ray data have not revealed evidence for pulsations and, hence, the identification is only based on positional evidence and must be treated with caution. The spectrum is hard and can be fitted by a power law with index -2.03 ± 0.26. However, if the pulsar identification is confirmed, it would indicate a new class of galactic gamma-ray sources, i.e. rotation-powered pulsars in binaries.

9.3.3 3EG J1835+5918: Geminga-like pulsar?

This is the strongest unidentified source at a high galactic latitude (25°) and is a good candidate to be a Gould Belt pulsar. EGRET observations give a spectral index between 70 MeV and 4 GeV of -1.7; it may steepen beyond that. There is no evidence for variability. An examination of all the x-rays sources in the error box reveals one object, RX1836.2+5925 (the brightest source in x-rays), which has no optical or radio counterpart [9]. Its x-ray spectrum shows two components: a soft thermal component and a hard component. There is no optical counterpart down to a limit of $V > 28.5$ [4]. No radio emission is observed down to a low level of emissivity. The large ratio of x-ray to optical luminosity and the emission of 100 MeV gamma rays is reminiscent of Geminga and it is suggested that this is also a radio-quiet pulsar, perhaps at a larger distance ($\sim$800 pc). Its gamma-ray luminosity, L_γ is 3.8×10^{34} erg s^{-1}, assuming the emission is isotropic and it is at this distance.

9.3.4 Galactic center

The galactic center is a region of intense interest and it is significant that there is an EGRET source, 3EG J1744-3039, positionally coincident with it. The galactic center ($l = 0°$, $b = 0°$) lies within the error circle of 0.2° radius and no gamma-ray source of comparable magnitude occurs within 15° of the center. The source has an unusual hard power-law spectrum (spectral index $= -1.3$) that steepens above 2 GeV (spectral index $= -3.1$). There is no evidence for variability nor is there any correlation with radio features, e.g. the CO distribution near the galactic center.

There does not seem much doubt that the source can be identified with the galactic center; the uncertainty concerns the nature of the object that emits the gamma rays. It is certainly unique, at least in the Galaxy. There have been several suggestions for the origin of the gamma-ray emission but none of them have been universally accepted. These include:

(a) a peak in the diffuse gamma-ray emission from the Galaxy;
(b) emission from a massive black hole as found near the center of the Galaxy;
(c) a young radio-quiet pulsar or pulsars, possibly in the foreground;
(d) a supernova remnant;
(e) annihilation from weak interacting massive particle (wimp) accumulation (neutralino) from dark matter in a gravitational cusp.

9.4 Microquasars

The gamma rays associated with blazars originate in jets of relativistic particles which emit synchrotron radiation (see appendix). The bulk motion associated with the jets is a large fraction of the velocity of light. The discovery of similar jets associated with galactic objects immediately raises the possibility that they might also be sources of gamma rays. Relativistic outflows have been reported from a variety of galactic objects (supernova remnants, pulsars, gamma-ray bursters) but the objects that seem most promising as gamma-ray sources are those associated with x-ray binaries.

The best known 'microquasars', GRS 1915+105 and 1E1740.7-2942, have long been regarded as good candidate gamma-ray sources [11] but they were not seen by EGRET. The velocity of the ejecta in GRS 1915+105 is as much as $0.98c$. Twelve sources are now listed as microquasars including Cygnus X-3, SS433, and Circinus X-1.

Perhaps the best studied object is SS433; it exhibits complex behavior which is clearly caused by beamed emission. However, although this is a powerful source of radio emission, it is only a weak x-ray source and is not detected at all at gamma-ray energies. As with blazars, it could be that the gamma-ray emission is strongly beamed and only those objects with beams pointing in our direction will be detectable. The best chance of detection at other wavelengths is with very high-resolution imaging and this is most effective when the beam is viewed at right angles.

Recently a new microquasar was discovered relatively nearby in radio studies with the VLBI and VLA [11]. Although not spectacular at optical or x-ray wavelengths, this x-ray binary, LS5039, shows bi-polar jets and is positionally coincident with the unidentified EGRET source, 3EG J1824-1514. It is also an x-ray source. At an assumed distance of 2–3 kpc, the gamma-ray luminosity (assuming isotropic emission) is $L_\gamma(> 100 \text{ MeV}) \approx 3.8 \times 10^{35}$ erg s^{-1}, considerably larger than the x-ray luminosity L_x (1.5–12 keV) $\approx 5 \times 10^{34}$ erg s^{-1}. The suggested mechanism for emission is inverse Compton scattering by the radio-synchrotron electrons on UV photons. If protons are also accelerated in the jet, the relativistic mass flow into the jet could be approximately 10^{37} erg s^{-1}.

Table 9.2. Unidentified sources from the Whipple survey.

Source	RA (1950) (hr min)	Dec. (1950) (deg arc-min)	Threshold (GeV)	Flux ($\times 10^{-10} \gamma$ cm^{-2} s^{-1})
Whipple-A	19 38	+30 04	130	20
Whipple-C	06 01	+43 08	100	11
Whipple-D	06 37	+05 54	210	7.7

The possibility that there are many other, as yet, undiscovered microquasars cannot be discounted; if pointing towards us, they would be very difficult to identify as microquasars but, by analogy with AGN, they would be the best candidates for gamma-ray detection.

9.5 VHE observations

Most VHE searches have concentrated on the positions of known or suspected candidate objects. This is particularly true of atmospheric Cherenkov detectors which have limited fields of view. The first complete survey (of the northern sky) was made in 1974–76 using the Whipple Observatory 10 m telescope which was multiplexed to give ten independent beams, each of 1°. The drift-scan mode of operation was employed so that a strip of sky 4° wide was scanned each night. Since these were first-generation non-imaging detectors, the sensitivity was limited. An upper limit ($<1 \times 10^{-9}$ photons cm^{-2} s^{-1}) was set for the region of sky from declination $-10°$ to $+70°$ with an energy threshold from 100 to 250 GeV [14]. Three regions gave signals of possible statistical significance; subsequent observations with greater sensitivity (two years later) did not confirm these sources whose coordinates are listed in table 9.2.

All three regions are within 10° of the galactic plane but are not obviously associated with any prominent objects. It is of interest that the Whipple-D lies close to 3EG J0634+0521 (as previously discussed in section 9.3.2).

The HEGRA group reported the detection of an unidentified source in the Cygnus region [15] which may be coincident with the EGRET source 3EG J2033+4118. A deep observation (113 hours) was obtained in an attempt to detect Cygnus X-3 which is nearby but outside the error circle. Although weak, the detection is statistically significant and augurs well for the next generation of VHE telescopes.

The only sky survey by a sensitive second-generation ACT was carried out by the HEGRA group who concentrated their observations on a narrow strip of the galactic plane from the galactic center to the Cygnus region (approximately one-quarter of the plane). This is a region with many interesting objects. No evidence was found for emission from any object with intensity greater than 20% of the Crab [16].

Air shower arrays are more suitable than ACTs for sky surveys but can only achieve good sensitivity by averaging over many days of observation. Hence, they can detect transient sources if they are very bright. In 1997–98 the Milagrito experiment operated for 16 months and set an upper limit to steady sources in the declination strip from 0° to 80° of $1-3 \times 10^{-10}$ photons cm^{-2} s^{-1} with energy threshold from 3 to 7 TeV [13]. Later, the Milagro experiment made a sky survey of the northern sky and found no sources brighter than twice the Crab Nebula.

Historical note: Geminga

The story of the identification of Geminga with a radio-quiet pulsar is a classical tale of astronomical detective work [12]. First discovered by the SAS-2 mission in 1973 and confirmed by COS-B, CG195+4 (as it was known in the COS-B catalog) was initially identified with an OB association and an SNR. The ratio of gamma-ray-to-radio intensity was noted as similar to that of the Crab and Vela pulsars: could it be a previously undiscovered pulsar? The discovery of an unusual x-ray source in the error box led to its identification with a faint optical counterpart. The total spectral energy distribution suggested that it might be a radio-quiet pulsar. A timing analysis of the x-ray data led to the discovery of a pulsar period which was confirmed in the EGRET data in 1992. Analysis of the archival data from SAS-2 and COS-B confirmed the detection. Nearly 20 years had passed between the initial discovery and the final identification which required the combined efforts of radio, optical, x-ray, and gamma-ray astronomers. With hindsight, it was realized that a blind period search in the COS B database would have found the periodicity in the gamma-ray signal.

Reference

[1] Carraminana A, Reimer O and Thompson D J 2001 *Proc. Workshop The Nature of Unidentified Galactic high energy Gamma-Ray Sources (Tonantzintla, Mexico, October 2000)* (Dordrecht: Kluwer)
[2] Gehrels N *et al* 2000 *Nature* **404** 63
[3] Grenier I A 2000 *GeV–TeV Gamma Ray Astrophysics Workshop (Snowbird, Utah, August 1999) (AIP Conf. Proc. 515)* ed B L Dingus, M H Salamon and D B Kieda (New York: AIP) p 261
[4] Halpern J P *et al* 2002 *Astrophys. J. Lett.* **573** L41
[5] Hartman R C *et al* 1999 *Astrophys. J. Suppl.* **123** 79
[6] Kaaret P 2001 *Proc. Workshop The Nature of Unidentified Galactic High Energy Gamma-ray Sources (Tonantzintla, Mexico, October 2000)* (Dordrecht: Kluwer) p 191
[7] Lamb R C and Macomb D J 1997 *Astrophys. J.* **488** 872
[8] Mayer-Hasselwander H A *et al* 1998 *Astron. Astrophys.* **335** 161

[9] Mirabel N *et al* 2000 *Astrophys. J.* **541** 180
[10] Reimer O 2001 *Proc. Workshop The Nature of Unidentified Galactic High Energy Gamma-ray Sources (Tonantzintla, Mexico, October 2000)* (Dordrecht: Kluwer) p 17
[11] Rodriguez L F and Mirabel F 2001 *Proc. Workshop The Nature of Unidentified Galactic High Energy Gamma-Ray Sources (Tonantzintla, Mexico, October 2000)* (Dordrecht: Kluwer) p 245
[12] Thompson D J 2001 *Proc. Workshop The Nature of Unidentified Galactic High Energy Gamma-Ray Sources (Tonantzintla, Mexico, October 2000)* (Dordrecht: Kluwer) p 3
[13] Wang K *et al* 2001 *Astrophys. J.* **558** 477
[14] Weekes T C 1988 *Phys. Rep.* **160** 1
[15] Aharonian F A *et al* 2002 *Astron. Astrophys.* **393** L37
[16] Aharonian F A *et al* 2002 *Astron. Astrophys.* **395** 803

Chapter 10

Extragalactic sources

10.1 Introduction

One of the major surprises from the EGRET observations on the Compton Gamma Ray Telescope was the large number of extragalactic sources discovered. Although the sky had been surveyed by the earlier SAS-2 and COS-B missions and many of the EGRET sources were of sufficient intensity to have been detected by them, the only extragalactic high energy gamma-ray source known prior to the launch of CGRO was the nearby quasar, 3C273 [10]. COS-B detected 3C273 but the observed spectrum seemed to steepen above 100 MeV and no time variability was observed. Hence, the COS-B team choose to concentrate their observing program on what seemed to be more interesting, the galactic plane. In so doing, they missed out on the very exciting field of extragalactic gamma-ray astronomy. In recent years, HE gamma rays have come to play an important role in the study of AGN and, potentially, in other extragalactic systems.

10.2 Galaxies: classification

Stars, the fundamental building blocks of the universe, are social creatures that do not like to be alone. Although the spacing between stars is much larger than their diameters, many stars have companions in a binary system. In some cases a planetary system may be a substitute for a companion star. Stars can also be found in clusters of as many as 10^5 stars. However, the most common star system is the galaxy in which there can be from 10^8 to 10^{12} stars. The Galaxy, or Milky Way, is a typical large galaxy with 10^{11} stars in a disk-like structure (chapter 4). On a larger scale, galaxies themselves occur in clusters (with relatively small spacing between them) and the clusters are members of so-called superclusters.

Galaxies are fundamental building blocks (like stars) and come in many varieties. The majority fit within the broad classification of ellipticals, spirals, and irregulars. As their name suggests, the ellipticals have a population of stars that form an ellipsoid; they contain little gas. The spirals have a disk-like structure

Table 10.1. EGRET observations of normal galaxies.

Galaxy	Type	Distance (kpc)	Predicted $F_\gamma(> 100\text{ MeV})$ ($\times 10^{-7}$ cm^{-2} s^{-1})	Observed $F_\gamma(> 100\text{ MeV})$ ($\times 10^{-7}$ cm^{-2} s^{-1})
LMC	Irregular	52	2.0 ± 0.4	1.9 ± 1.4
SMC	Irregular	54	2.4	<0.5
Andromeda	Spiral Sb	570	0.2	<0.5

in which the stars have a distribution that looks like it contains many open spiral arms. The irregulars are exactly as their name suggests. Both the spirals and irregulars contain significant amounts of gas and dust. On average, spirals tend to be larger than ellipticals which are larger than irregulars but the scales overlap. However, at the center of clusters there is often a very large elliptical galaxy.

Within clusters, galaxies interact and sometimes collide. Because of the large spacing between stars compared to their diameters, the galactic collisions seldom involve stellar collisions. However, these galactic encounters do result in significant gravitational distortions and are probably important in determining the evolution of the galaxies. Despite intensive study, there is still no clear consensus as to the evolution of galaxies.

A small proportion of the overall population of galaxies falls under the heading of Active Galaxies which are discussed in section 10.5.

10.3 Normal galaxies

Since the strongest source in the 100 MeV sky is the galactic plane, it is natural to look to other galaxies as potential gamma-ray sources. The Galaxy is a normal galaxy in the sense that it is not ultra-luminous like Seyferts, blazars, or radio galaxies (see later) and the bulk of the observed optical radiation seems to be the sum total of all the stellar emission. The Galaxy is an Sb spiral and is one of the largest members of a small group of galaxies, the Local Group, which has about 20 members. The Local Group is a member of the Virgo Cluster with about 10^3 galaxies. On a larger scale, the Virgo Cluster is a member of the Virgo Supercluster whose membership may exceed 10^8 galaxies.

The closest large normal galaxy is M31, the Andromeda Nebula (distance = 570 kpc). It is also an Sb spiral and appears to be very similar to the Galaxy. In fact, we deduce many of the properties of the Galaxy from the study of M31. Because of the similarity we can, with some confidence, predict the gamma-ray emission level of M31 by analogy with the galactic emission. This turns out to be less than the minimum sensitivity of EGRET so it is no surprise that it has not been detected (table 10.1).

The Small Magellanic Cloud (SMC) is a small irregular galaxy (distance = 54 kpc). The detection of the SMC in 100 MeV gamma rays was regarded as a potential way of demonstrating that the cosmic ray density seen here in the Solar System was of extragalactic origin [3]. However, only an upper limit was obtained from EGRET observations (see historical note: cosmic ray origins) [7] (table 10.1).

The Large Magellanic Cloud (LMC) is also a nearby small irregular galaxy (distance = 52 kpc). Because it is irregular, it is more difficult to estimate its cosmic ray production rate and gas content. It is marginally detected by EGRET [6] (table 10.1).

The galactic plane has not been seen at VHE energies (chapter 4) and neither has any other normal galaxy.

10.4 Starburst galaxies

Starburst galaxies are so called because they seem to involve a very high rate of star formation. They are identified by their high luminosity in the infrared and their extended emission regions in the radio and x-ray bands. There is a high rate of star formation near the core of the galaxy. They also have a much higher rate of supernova explosions (about ten times higher than that of normal galaxies). The infrared emission comes from dust in the hot interstellar medium where the stars are forming. The enhanced rate of supernova explosions indicates that the cosmic ray density may be greater and, hence, that they might be detectable gamma-ray sources. However, the two nearest starburst galaxies, M82 and NGC 253, have thus far yielded only upper limits at HE energies. They are seen in hard x-rays with the Oriented Scintillation Spectroscopy Experiment (OSSE). The starburst emission may only be a phase but one lasting in excess of 20 million years.

The CANGAROO group have reported the detection of a strong signal from NGC 253 at energies in excess of 500 GeV [4]. The detection of this new class of object is strengthened by the apparent finite extent of the source which matches that of the visible galaxy.

10.5 Active galaxies

The taxonomy of extragalactic sources is complex and the sub-divisions are more historical than physical. As more of their properties are measured and their spectral energy distribution (SED) determined over a wide range of energies, it becomes possible to classify them more meaningfully. By definition, active galaxies are those whose core luminosity exceeds the norm and where there is evidence for relativistic particle acceleration.

Historically, the first active galaxies noted were spiral galaxies which appeared to have bright cores and broad emission lines. Later these came to

Active galaxies 129

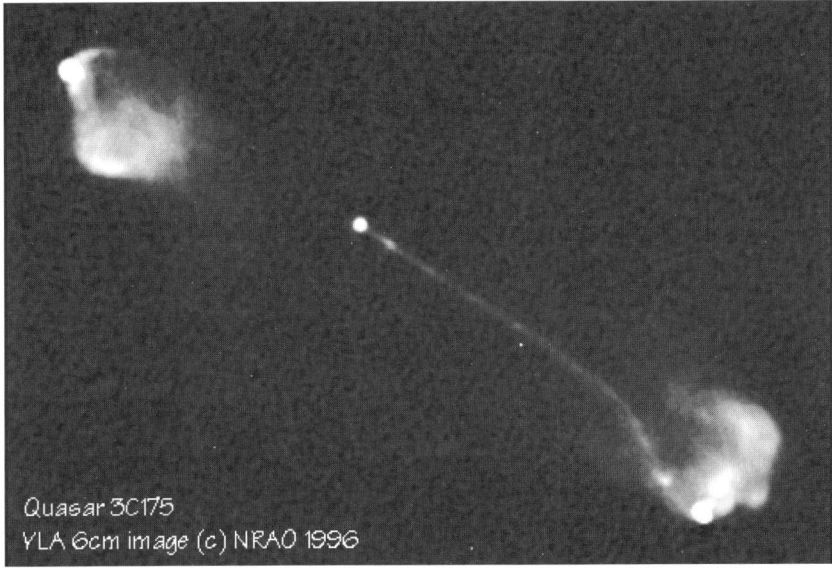

Figure 10.1. The radio image of the radio galaxy, 3C175 as recorded by the Very Large Array (VLA) at 6 cm wavelength. The central core (black hole) is clearly visible as are the radio blobs at each end of the jets (only one of which is visible here). (Figure: NRAO/NSF.)

be called Seyfert galaxies after their discoverer, Carl Seyfert. They are a sub-classification of a more general class known as active galactic nuclei (AGN). These are now generally defined so as to include radio galaxies, quasars, quasi-stellar objects (QSOs), and blazars.

10.5.1 Radio galaxies

The classification 'radio galaxy' is generally used to describe the most energetic nearby galaxies; it is somewhat obsolete since these are really nearby AGN and not a seperate class. They are thus prime targets for gamma-ray studies and were the subject of the earliest optimistic predictions of gamma-ray emission from extragalactic objects. Their radio luminosity is some thousand times greater than a normal galaxy, they are observed to have spectacular 'lobes' of radio emission far outside the visible galaxy, and narrow jets which seem to connect the center of the galaxy to the lobes. They are clearly host to relativistic particles. Classic examples are Cygnus A and Centaurus A (figure 10.1). If cosmic rays have an extragalactic origin, these objects would be the prime candidates for the sources because the radio emission requires the presence of large numbers of highly relativistic electrons. Because they are so close these objects can be directly imaged and resolved at radio, optical, and x-ray wavelengths so their properties are well known.

Most interest has centered on Cen A because it is the closest ($z = 0.0018$) and brightest. The radio image shows structure on many scales. The most prominent features are the broad lobes which extend over a few degrees. But there are many other features: inner lobes, connecting x-ray jets, an active nucleus, dust lanes. It may well be a blazar that makes a large angle (60–75°) with our line-of-sight but it is generally treated somewhat separately from blazars. Gamma-ray emission from Cen A was first detected at VHE energies by an experiment in Australia in 1975 [5]. The observed flux was $(4.4 \pm 1.0) \times 10^{-11}$ photons cm^{-2} s^{-1} for $E > 300$ GeV. However, the detection was not confirmed by later, more sensitive, observations. These observations could be explained in terms of a Compton-synchrotron model in which synchrotron microwave photons were the target photons for gamma-ray production. In 1975, the source was very bright at microwave wavelengths and it is possible that the gamma-ray source is time variable; it has not achieved this microwave brightness since 1975.

Although it appears to have been in a low state throughout the duration of the CGRO mission, it was, in fact, detected by all four instruments on CGRO (BATSE, OSSE, COMPTEL, and EGRET), the only extragalactic object to be so detected [9]. The observed gamma-ray spectrum is complex and it cannot be fitted by a thermal spectrum or a single power law. The EGRET source, 3EG J1324-4314, is time variable and has a spectral index of 2.58 ± 0.26. The weak detection of Cen A by EGRET is the only HE detection of a radio galaxy.

10.5.2 Active galactic nuclei

Since their discovery in the 1960s, quasars have dominated extragalactic astronomy by virtue of their great luminosities. It is now generally accepted that the power source of quasars are supermassive black holes of mass 10^{8-9} M$_\odot$. Some 10% of all quasars are more luminous at radio wavelengths than at optical ones and are, hence, called radio-loud. The radio emission is believed to originate in the associated jets which are aligned with the poles of the spinning black hole. The large radio lobes are sometimes seen to occur near the ends of the jets. The jets are a key feature of many of these objects but is not clear why they arise. Since they do not always occur, they may be associated with only the more massive black holes or with their rotation. The key features in the canonical model are shown in figure 10.2. The main components are the supermassive black hole, the hot accretion disk surrounding it, the dust torus in the same plane as the disk but larger, the emission line gas clouds that are distributed somewhat randomly about the black hole, the two relativistic jets that emerge perpendicular to the accretion disk, and the large lobes where the jets terminate.

It is generally believed that the jets channel a plasma flowing out with relativistic speed, and any radiation produced inside them is received greatly

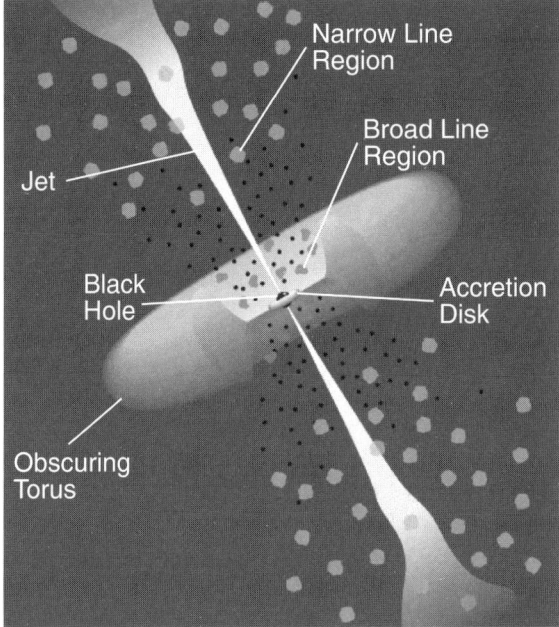

Figure 10.2. Cartoon of an active galaxy.

modified by the Doppler effect. If the jet aims straight towards us, the radiating regions can almost keep up with the radiation they emit; seen by us, both the waves and the durations of outbursts are compressed by a large Doppler factor, δ. This greatly enhances the power we receive when we are 'looking down the gun barrel' and, when we also allow for the angular beaming, the power received on a detector is increased by such a large factor, δ^4, that it outshines everything else from the galaxy [1].

The unified theory of AGN suggests that the wide variety of AGN types seen is largely a function of viewing angle and, hence, of geometry rather than physics. It is generally believed that the jets are filled with relativistic particles and that much of the observed radio emission is synchrotron radiation from these particles. A thick torus of dust surrounds the accretion disk and may obscure it completely if the system is viewed edge-on. The radio lobes are then clearly seen and the AGN is seen as a radio galaxy with the radio emission dominant. The central core may be almost completely obscured by the dust torus.

When the jets make an acute angle with the line of sight, the radio and optical emission from the core can be seen. Broad emission lines from the gas clouds are also visible. Such objects are classified as lobe-dominated quasars. As the

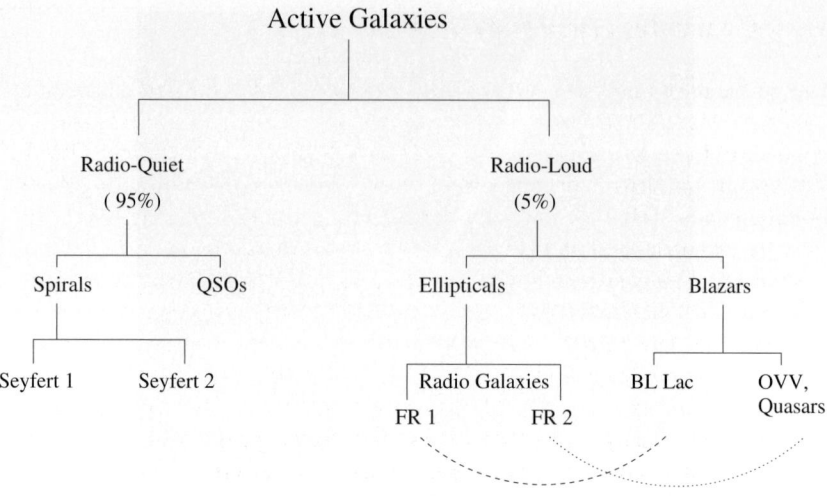

Figure 10.3. Classification scheme for active galaxies [2].

viewing angle decreases, the core becomes more apparent and the objects are classified as core-dominated quasars. The synchrotron emission from the jet also becomes the strongest emission mechanism and is both variable and polarized. Eventually, when the viewing angle is very small, the jet is the most obvious feature and the object is called a blazar. The extreme blazar is one in which the viewing angle is zero, i.e. the observer is looking straight down the jet. If no emission lines are seen, the object is classified as a BL Lacerate (BL Lac) object (named after the prototype source with these properties). If the blazar does exhibit emission lines, it is classified as a Flat Spectrum Radio Quasar (FSRQ). The viewing situation in which the observer is almost on the jet axis has been compared to looking straight down a gun barrel or into a particle accelerator beam

BL Lacs are difficult to study optically because of the absence of emission lines; hence, distances are difficult to establish. They also exhibit more polarization than other AGN. BL Lacs were originally subdivided according to whether they were discovered in radio surveys, i.e. radio-selected BL Lacs (RBLs) or in x-ray surveys, i.e. x-ray-selected BL Lacs (XBLs). Recently these subdivisions have been generalized to Low-frequency BL Lacs (LBLs) and High-frequency BL Lacs (HBLs). We shall see in the next two chapters that these classifications have some physical meaning and are strong predictors of their gamma-ray emitting properties.

One classification system of active galaxies [2] is shown in figure 10.3.

Historical note: cosmic ray origins

One of the most important early results from EGRET was the non-detection of the Small Magellanic Cloud [3]. Because of its proximity, a good estimate can be made of its target hadron content in the same way as was done for the Galaxy. The cosmic ray density depends on whether the cosmic radiation is of galactic or extragalactic origin. Based on the observed synchrotron radio emission, the galactic magnetic field and the electron-to-proton ratio, the internal (galactic) cosmic ray density was estimated. If the galaxy is tidally disrupted (perhaps from an encounter with the Large Magellanic Cloud), then the density will be somewhat less than if it is in a quasi-static equilibrium situation. The observed integral flux was ($F_\gamma(>100\text{ MeV})<0.5 \times 10^{-7}\text{ cm}^{-2}\text{ s}^{-1}$ (table 10.1). The predicted flux from an extragalactic density of cosmic rays, similar to that observed locally, would be almost five times greater than this [3, 8]; this discrepancy would seem to eliminate the extragalactic origin theory and point to a local (galactic) origin for the cosmic radiation, at least up to VHE energies. The observed upper limit is also incompatible with a quasi-static galactic production model in the SMC and confirms the hypothesis that the galaxy is tidally disrupted.

References

[1] Buckley J 1998 *Science* **279** 676
[2] Dermer C D 1994 *Proc. NATO Advanced Study Institute 'The Gamma-Ray Sky'* (Dordrecht: Kluwer Academic) p 39
[3] Ginzburg V L 1972 *Nature* **239** 8
[4] Itoh C *et al* 2002 *Astron. Astrophys.* **396** L1
[5] Grindlay J *et al* 1996 *Astrophys. J. Lett.* **1559** 100
[6] Sreekumar P *et al* 1992 *Astrophys. J. Lett.* **400** L67
[7] Sreekumar P *et al* 1993 *Phys. Rev. Lett.* **70** 127
[8] Sreekumar P and Fichtel C E 1991 *Astron. Astrophys.* **251** 447
[9] Steinle H *et al* 1997 *Proc. 4th Compton Symposium, (AIP Proc. 410)* ed C D Dermer, M S Strickman and J D Kurfess (New York: AIP) p 1298
[10] Swanenburg B N *et al* 1978 *Nature* **275** 298

Chapter 11

Active galactic nuclei: observations

11.1 Gamma-ray blazars

Although AGN come in a variety of forms, it is almost exclusively the blazar sub-classifications of FSRQs and BL Lacs that have thus far proven to be detectable at energies above 10 MeV. The term 'blazar' is not rigorously defined but is usually taken to include Optically Violently Variable quasars (OVV), BL Lacs, sources that are radio-loud with compact cores, and sources that exhibit superluminal motion or strong polarization. Very often a source fits into more than one of these categories. Clearly, these are non-thermal sources with the observation of polarization, a flat radio spectrum ($\alpha_r > -0.6$), x-ray emission, and a high degree of variability at all wavelengths.

Whereas VHE radiation has been detected almost exclusively from BL Lacs, HE emission comes from some BL Lacs but more usually from the FSRQs. It is thus appropriate to consider HE and VHE observations separately.

11.2 Gamma-ray observations: HE

11.2.1 HE source catalog

Almost all of our information on gamma-ray emission from blazars in the energy range 20 MeV to 10 GeV has come from EGRET observations. In the Third Catalog of EGRET observations [7], more than 70 AGN which emit gamma rays at energies above 100 MeV have been identified based on their positional coincidence with a previously established catalog of more than 400 flat-spectrum radio sources which had a radio power of more than 1.0 Jy at 5 GHz. A substantial fraction of the unidentified high-latitude sources in the EGRET Catalog are likely to be AGN as well; for a variety of reasons these might not have been included in the initial catalog of radio sources. The blazars are, by far, the largest class of identified objects in the catalog. Some of these extragalactic sources are also detected by COMPTEL and a few by ground-based telescopes.

These extragalactic gamma-ray sources have two remarkable properties: in the majority of them the gamma-ray luminosity dominates the power output of the blazar; and short-term variations are seen on times scales of hours to months and years [19]. The typical EGRET exposure was of one to two weeks duration and the threshold sensitivity at energies >100 MeV was approximately 3×10^{-7} photons cm^{-2} s^{-1}. The criteria for detection of an AGN was that the signal exceed 5σ if the source was within $10°$ of the galactic plane, 4σ elsewhere. The properties of ten of the most interesting HE blazars in the Third EGRET Catalog [7] are listed in table 11.1. The first column gives the EGRET Catalog designation. The second column is the designation of the source (if detected) in the GeV Catalog [13]. A non-detection indicates that the source is weak at energies >1 GeV. The third column gives the name of the AGN with which the source has been identified. The fourth column lists the value of θ_{95}, the angular radius of the 95% confidence contour; its size is some measure of the uncertainty in the identification. In practice, many of the identifications have been verified by correlated variations in intensity. The fifth column lists the range of average flux levels detected when averaged over the viewing period. The maximum flux levels on shorter time scales often exceeded these values. The sixth column gives the differential photon spectral index with its uncertainty. The last column lists the redshift.

11.2.2 Distance

The measured redshifts range from 0.03 to 2.28.

11.2.3 Classification

Most of the identified AGN are classified as flat-spectrum radio sources. A smaller number are identified with BL Lacs, including BL Lac itself. This contrasts with the TeV-emitting AGN where the sources are BL Lacs.

11.2.4 Time variability

One of the characteristic features of gamma-ray-emitting blazars is their time variability on scales ranging from hours to years. Variability is a surprising feature in some respects because it implies a small emission region. If low energy photons (e.g. infrared, optical, and ultraviolet) are produced in the same region, some gamma-ray photons will pair produce with these low energy photons and will not escape. If the variable emission originates near the base of the jet, there is considerable ambient radiation present which can attenuate the gamma-ray signal. This opacity problem is reduced considerably if the emission is beamed toward the observer; and this has been one of the main arguments for gamma-ray beaming in these objects. The blazar PKS1622-297 has shown the shortest doubling time (4 h). PKS0528+134, the AGN near the Crab Nebula in the sky, is one of the most

Table 11.1. Selected HE blazars [7].

EGRET 3rd Cat. 3 EG	GeV Cat. (GEV)	Identification	type	θ_{95} (deg)	F_γ (>100 MeV) ($\times 10^{-7}$ cm^{-2} s^{-1})	Index	Redshift z
J0222+4253	J0223+4254	3C66A		0.31	1.5–2.5	2.01±0.14	0.444
J0530+1323	J0530+1340	PKS0528+134		0.21	3.5–35.1	2.46±0.04	2.060
J0853+1941		OJ+287		0.91	1.0–1.6	2.03±0.35	0.306
J1104+3809	J1104+3809	Mrk421		0.21	0.9–2.7	1.57±0.15	0.031
J1222+2841	J1222+2837	W Comae		0.29	1.1–3.3	1.73±0.18	0.102
J1229+0210		3C273		0.32	0.8–4.8	2.58±0.09	0.158
J1255-0549	J1256-0546	3C279		0.08	0.8–26.7	1.96±0.04	0.538
J1329+1708		PKS1331+170		0.73	0.4–3.3	2.41±0.47	2.084
J2158-3023		PKS2155-304		0.68	0.8–3.0	2.35±0.26	0.116
J2202-4217		BL Lac		1.05	0.9–4.0	2.60±0.28	0.069

variable sources; it became brighter than the Crab in March, 1993. In 3C279, one of the best studied and closest blazars, the shortest doubling time observed was 8 h. It is probable that shorter variations will be seen with more sensitive instruments since the shortest variations seen to date are limited by the number of photons detected. Many other AGN show variations on time scales of days. In general, the EGRET-detected AGN that are identified with flat-spectrum radio sources are more variable than those identified with BL Lacs.

11.2.5 Luminosity

The absolute luminosity at MeV–GeV energies, assuming isotropic emission, ranges from 7×10^{44} to 4×10^{48} erg s^{-1}. The BL Lacs are closer and their absolute luminosities are at the low end of this range.

11.2.6 Spectrum

The EGRET experiment provided good measurements of sources from 30 MeV to 10 GeV. All of the measured spectra can be represented by a simple power law:

$$F(E) = k(E/E_0)^{-\alpha} \text{ photons cm}^{-2} \text{ s}^{-1} \text{ MeV}^{-1}$$

where the spectral index α and k are free parameters. The values of α vary from 1.6 to 2.8 with an average value of 2.2. The EGRET measurements do not show any indication of a cut-off up to the highest measured energy. However, VHE telescopes have detected only four of these blazars which indicates that there is significant spectral steepening between 10 and 300 GeV. The spectral indices do not show any consistent correlation with redshift.

It should be noted that the measured cosmic ray spectral index in this energy range is about -2.7 so that these are unexpectedly flat spectra. If the gamma-ray spectra are representative of the spectra of progenitor particles accelerated in the sources and if these are the sources of cosmic rays, then clearly the particle spectra steepen as they transit the intervening space. Some typical spectra are shown in figure 11.1.

11.2.7 Multi-wavelength observations

Some of the most interesting results on gamma-ray sources have come through observations where several telescopes operating at different wavelengths simultaneously monitor the activity of a blazar. These multi-wavelength campaigns have involved the larger astronomical community in the study of HE sources and served the subsidiary purpose of confirming the source identifications with the blazar in the error box. The best hope for understanding the mechanisms at work in blazars comes from the combination of observational data over the whole electromagnetic spectrum. Since such objects vary at all wavelengths, to be meaningful, the spectra should be measured simultaneously at each wavelength.

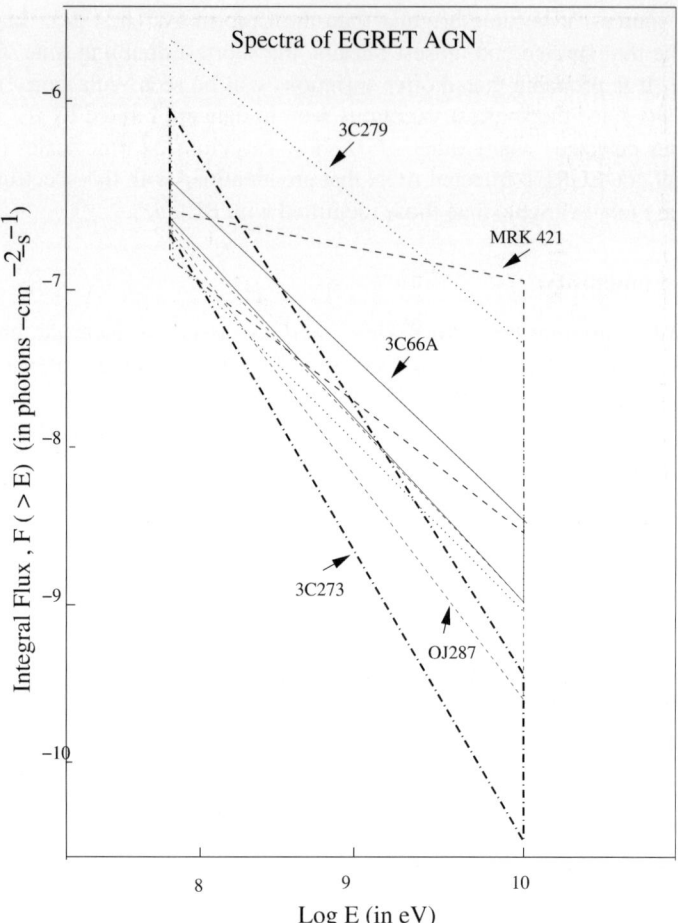

Figure 11.1. The integral spectra of five AGN detected by EGRET with their statistical uncertainties [7].

This is very difficult to realize in practice because of the different scheduling practices and priorities in the various wavebands. Nonetheless, significant efforts have been made to organize observing campaigns since the detection of the first blazar in a high state, 3C279 [9] (see historical note: discovery of 3C279). However, very often the spectra obtained are only quasi-simultaneous which makes interpretation difficult. Even 3C279, which is one of the most intensely studied blazars, is still poorly sampled. The spectra shown in figure 11.2 show that the percentage variability seen in 1996 was strongest in the MeV–GeV bands.

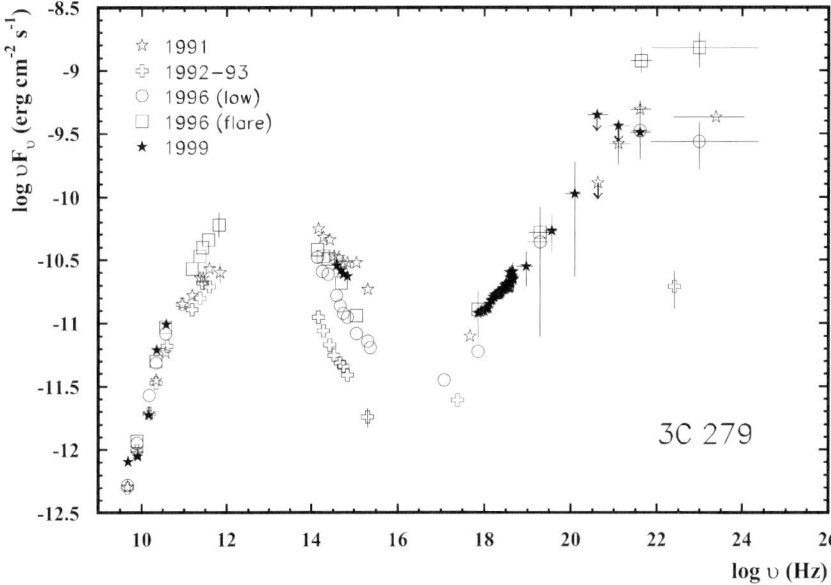

Figure 11.2. The spectral energy distribution of 3C279 at different times showing the large variations and high luminosity at MeV/GeV energies as observed by COMPTEL and EGRET [5]. (Figure: W Collmar.)

11.2.8 Spectral energy distributions

The spectral energy distribution (SED) is generally plotted as νF_ν against ν. This type of plot is often referred to as a power spectrum since it is a measure of the power observed at each frequency. The spectral energy emission of the HE blazars appears to consist of two parts. First, a low energy component exhibits a power per frequency distribution that rises smoothly from radio wavelengths up to a broad peak in the range spanning microwave to infrared wavelengths, depending on the specific blazar type, above which the power output rapidly drops off. Second, a distinct, high energy, component, which does not extend smoothly from the low energy component, is seen. Typically, it becomes apparent in the x-ray range and has a peak power output in the gamma-ray range between 1 MeV and 1 GeV. On a power plot the SED shows a two-humped shape which is often characteristic of an inverse-Compton synchrotron model. The SED may be complicated by the presence of the spectra of other, usually thermal, components of the AGN which are superimposed on the simple two-humped SED.

11.2.9 Future prospects

With the anticipated sensitivity of GLAST, the number of detected AGN at 100 MeV energies could increase a hundred-fold. With this large sample, it should be possible to make significant sub-divisions into different kinds of blazar and to understand them in terms of a unified theory. It will be possible to study the evolution of the gamma-ray-emitting AGN with time and to compare this with other evolutionary parameters. It may be possible to determine whether the observed extragalactic emission at 100 MeV energies is truly diffuse or whether it is the sum of all AGN (chapter 14). With increased sensitivity, shorter time variations will be measurable so that detailed comparisons may be made with variations in other wavebands. It should also be possible to detect other types of active galaxies. For relatively nearby radiogalaxies, such as Cen A, it will be possible to determine whether the emission region is truly at the core of the galaxy.

11.3 Gamma-ray observations: VHE

11.3.1 VHE source catalog

Ground-based observations of quasars were made as early as 1964 [14], shortly after their discovery and before their nature was known; in principle then, they might have been discovered as gamma-ray sources before the availability of large space platforms. However, the ground-based techniques available were relatively insensitive. No credible VHE detections were reported until AGN had been established as MeV–GeV sources by EGRET, by which time the sensitive atmospheric Cherenkov imaging technique had been developed. Given the high degree of variability now established in TeV sources, it is perhaps fortunate that the first source seen at VHE energies was the Crab Nebula, a notoriously steady source. One can only speculate at the controversy that would have ensued if the highly variable Markarian 421 (Mrk421) or Markarian 501 (Mrk501) had been the first source reported with the new VHE techniques!

The BL Lac object, Mrk421, at $z = 0.031$, was detected as the first extragalactic source of VHE gamma rays in 1992 by the Whipple Observatory gamma-ray telescope [16]. A two-dimensional image of Mrk421 in TeV gamma rays is shown in figure 11.3. The observations which led to the detection of Mrk421 at TeV energies were initiated in response to the detection of newly-discovered EGRET AGN. The initial detection indicated a 6σ excess and the flux above 500 GeV was approximately 30% of the flux of the Crab Nebula at those energies. Mrk421 was soon confirmed as a source of VHE gamma rays by several other ground-based experiments. The uncertainty in the source location at VHE energies (0.05°) was significantly less than at HE energies (0.5°) because Mrk421 is such a weak source at 100 MeV energies (figure 11.3). Subsequent multi-wavelength correlations have confirmed that the VHE source is indeed Mrk421 and not some other object in the error box.

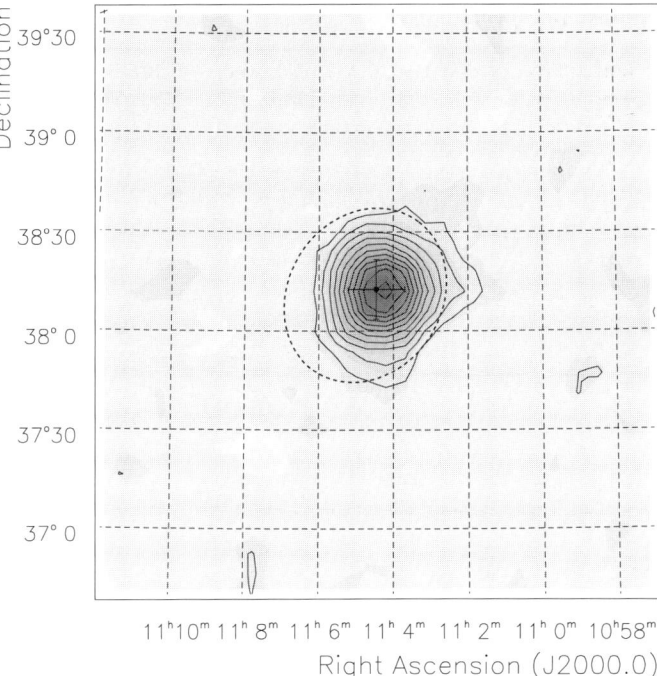

Figure 11.3. Two-dimensional plot of the VHE gamma-ray emission from the region around Mrk421. The gray scale is proportional to the number of excess gamma rays and the contours correspond to 2σ levels. The ellipse (full dashes) give the 95% confidence interval determined by EGRET [18]. The position of Mrk421 is indicated by the cross [2]. (Reproduced with permission from the *Astrophysical Journal*.)

Given the successful detection of Mrk421 and the flat spectra of the many EGRET AGN, it might be expected that many more would be detectable at VHE energies. This has not proven to be so—even prominent objects, like 3C273 and 3C279, have spectra that steepen in the 10–100 GeV range. By looking at BL Lacs that were similar to Mrk421, but which had not been detected by EGRET, the second TeV-emitting BL Lac object Mrk501 ($z = 0.034$) was detected [17]. Although Mrk501 was subsequently seen by EGRET when it flared, this VHE result was significant because it was the first gamma-ray source to be discovered by ground-based observations. Hence, VHE extragalactic gamma-ray astronomy was established as a legitimate channel of astronomical investigations in its own right, not just an adjunct of high energy observations from space. The flux of Mrk501 during 1995 was, on average, 10% of the VHE flux of the Crab Nebula, making it the weakest detected source of VHE gamma rays detected up to then.

In all, eight extragalactic objects have been reported as sources of VHE

Table 11.2. VHE blazars [20].

Source	Type	z	Discovery date	EGRET third Cat.	Grade
Markarian 421	XBL	0.031	1992	yes	A
Markarian 501	XBL	0.034	1995	no	A
1ES2344+514	XBL	0.044	1997	no	B
1ES1959+650	XBL	0.048	1999	no	A
BL Lac	RBL	0.069	2001	yes	C
PKS2155-304	XBL	0.116	1999	yes	B
H1426+428	XBL	0.129	2000	no	A
3C66A	RBL	0.44	1998	yes	C

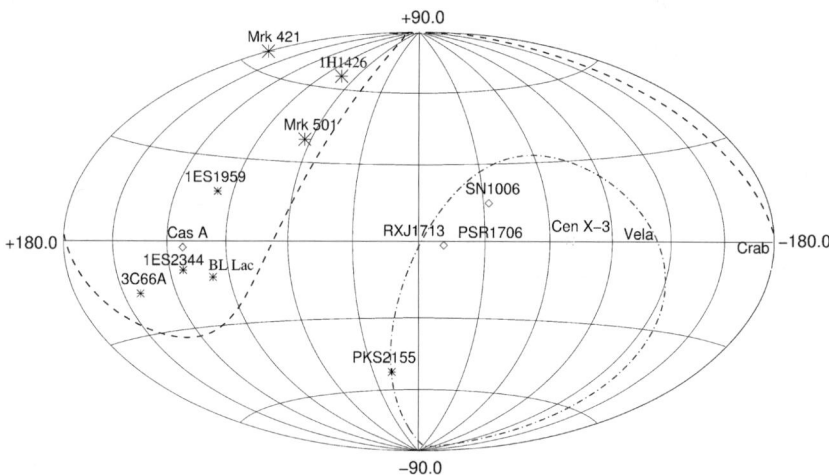

Figure 11.4. Distribution of TeV sources in sky. Note that unlike the corresponding map for 100 MeV sources (figure 9.1), all the sources are identified.

gamma rays and their properties are summarized in table 11.2 [20]. Their distribution in the sky is shown in figure 11.4. It should be noted that only four of the AGN entries have been independently confirmed at the 5σ level (but this also holds from most of the EGRET sources) and, hence, are classified as A sources. The allotted grade gives some measure of the credibility that should be assigned to the reported detections. On this scale the EGRET sources would be classified as B (except for 3C273 which would be classified as A since it was also detected by COS-B).

11.3.2 Distance

All of the VHE blazars detected to date are relatively closeby with redshifts ranging from 0.031 to 0.129, and perhaps to 0.44. The latter detection is somewhat suspect: the signal is not strong, the source is classified as an LBL, it is reported by only one group, and it is not expected that gamma rays with $E_\gamma \approx$ 1 TeV would be detected from a source at this distance because of absorption by pair production on the extragalactic infrared background (see chapter 14).

11.3.3 Classification

The well-established VHE sources, Mrk421, Mrk501, 1ES1959+650, and H1426+428 are classified as HBLs. As we shall see in the next chapter, the VHE sources are easiest to understand as HBL sources.

11.3.4 Variability

Extreme variability on time scales from minutes to years is the most distinctive feature of the VHE emission from these BL Lac objects. The first clear detection of flaring activity in the VHE emission of an AGN came in the 1994 observations of Mrk421 by the Whipple telescope where a 10-fold increase in the flux, from an average level that year of approximately 15% of the Crab flux to approximately 150% of the Crab flux, was observed. The observations of Mrk421 in 1995 [2], shown in figure 11.5, revealed several distinct episodes of flaring activity as in previous observations; perhaps more importantly, they indicated that the VHE emission from Mrk421 was best characterized by a succession of day-scale or shorter flares with a baseline emission level below the sensitivity limit of the Whipple detector.

The hypothesis that the VHE emission from Mrk421 could flare on sub-day time scales was borne out in spectacular fashion in 1996, with the observations of two short flares [6] (figure 11.6). In the first flare, observed on 7 May, the flux increased monotonically during the course of ∼2 hr of observations. This flux is the highest observed from any VHE source to date. The doubling time of the flare was ∼1 hr. The next night the flux had dropped to a flux level of ∼30% of the Crab Nebula flux, implying a decay time scale of <1 day. The second flare, observed on 15 May, although weaker, was remarkable for its very short duration: the entire flare lasted approximately 30 min with a doubling and decay time of less than 15 min. These two flares exhibited the fastest time scale variability, by far, seen from any blazar at any gamma-ray energy.

In 2001, Mrk421 underwent an extraordinary period of activity during which it was consistently brighter than the Crab Nebula over a three-month period. Several bright flares were observed.

In 1997, the VHE emission from Mrk501 increased dramatically. After being the weakest known source in the VHE sky in 1995–96, it became the brightest, with an average flux greater than that of the Crab Nebula (whereas

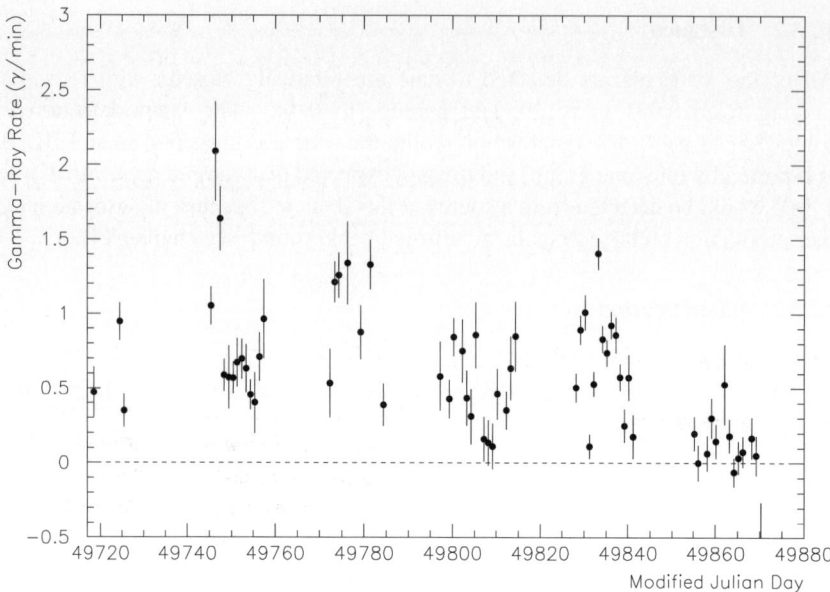

Figure 11.5. Daily VHE gamma-ray count rates for Mrk421 during 1995 as recorded at the Whipple Observatory. Modified Julian Day 49 720 corresponds to 3 January 1995 [2]. (Reproduced with permission from the *Astrophysical Journal*.)

previous observations had never revealed a flux >50% of the Crab flux). Also, the amount of day-scale flaring increased and, for the first time, significant hour-scale variations were seen. The six-month history of observations by the HEGRA telescope [10] is shown in figure 11.7.

1ES1959+650 and 1ES2344+514 also show variability on a time scale of hours–days. It appears likely that the other VHE blazars are also variable but, because they are weaker, their variability is more difficult to detect.

11.3.5 Luminosity

Blazars, whose gamma-ray emission peaks at HE energies, are generally more luminous than those that peak at VHE energies. This has led to the suggestion that there is an anti-correlation with peak gamma-ray energy and gamma-ray luminosity. Even when the VHE emission is very strong, it does not exceed the x-ray luminosity at that time.

11.3.6 Spectrum

Accurate measurements of the VHE spectrum are important for a variety of reasons. First, the shape of the high energy spectrum is a key input parameter of

Gamma-ray observations: VHE

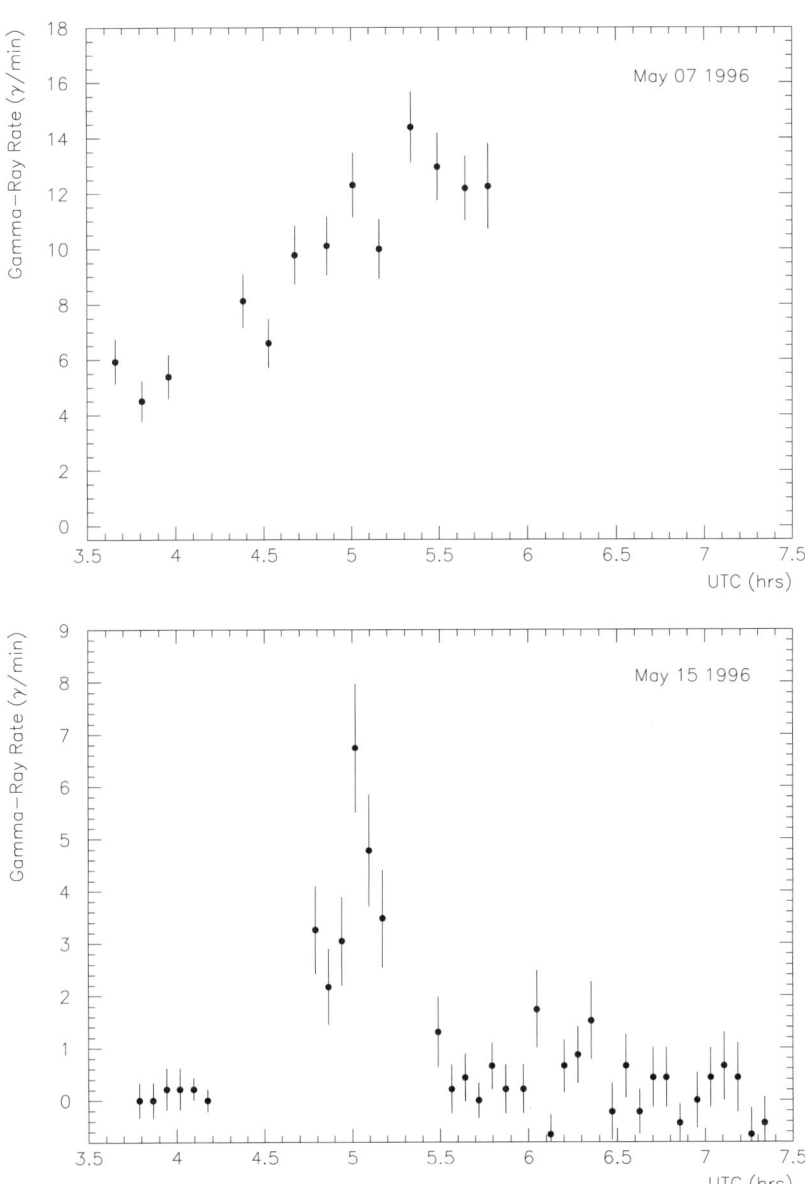

Figure 11.6. Light curves of two flares observed from Mrk421 by the Whipple Collaboration on 7 May 1996 (*a*) and 15 May 1996 (*b*). The time axes are shown in coordinated universal time (UTC) in hours. For the 7 May flare, each point is a 9-min integration; for the 15 May flare, the integration time is 4.5 min [6].

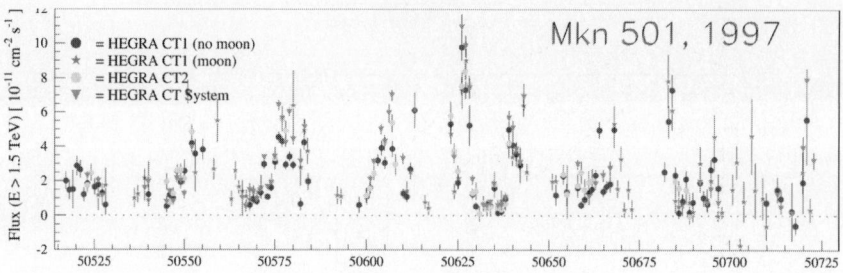

Figure 11.7. Nightly averages of the TeV gamma-ray flux from Mrk501 as recorded by the HEGRA experiment. For this extended campaign of observations, the telescope was operated in bright moon time [10].

AGN emission models. Second, the way in which the spectrum varies with flux, compared to longer wavelength observations, provide further emission model tests. Third, spectral features, such as breaks or cut-offs, can indicate changes in the primary particle distribution or absorption of the gamma rays via pair-production with low energy photons at the source or in intergalactic space.

The high-flux VHE emission from Mrk501 in 1997 and Mrk421 in 2001 has permitted detailed spectra to be extracted. Measurements are possible over nearly two decades of energy. As many as 25 000 photons were detected in these outbursts so that the spectra were derived with high statistical accuracy. Unlike the HE sources where the photon-limited blazar measurements are consistent with a simple power law, there is definite structure seen in the VHE measurements. The spectra of Mrk421 and Mrk501 can each be fit with a spectrum with an exponential cut-off (figure 11.8(*a*)).

The Mrk421 spectrum can be represented by

$$\frac{dN}{dE} \propto E^{-2.14\pm0.03_{\text{stat}}} \exp\left[-\frac{E}{4.3 \pm 0.3_{\text{stat}}(-1.4 + 1.7)_{\text{syst}}}\right]$$

m^{-2} s^{-1} TeV^{-1} [11] and the Mrk501 spectrum by

$$\frac{dN}{dE} \propto E^{-1.92\pm0.03_{\text{stat}}\pm0.20_{\text{syst}}} \exp\left[-\frac{E}{6.2 \pm 0.4_{\text{stat}}(-1.5 + 2.9)_{\text{syst}}}\right]$$

m^{-2} s^{-1} TeV^{-1} [1] where E is in units of TeV.

There is definite evidence for a change in the shape of the spectrum of Mrk421 as a function of the total luminosity of the source. However, all the spectra are consistent with the same cut-off energy (figure 11.8(*b*)).

For Mrk421, the exponential cut-off energy is approximately 4 TeV and for Mrk501 it is approximately 3–6 TeV. The coincidence of these two values suggests a common origin, i.e. a cut-off in the acceleration mechanisms with the blazars or perhaps the effect of the infrared absorption in extragalactic space.

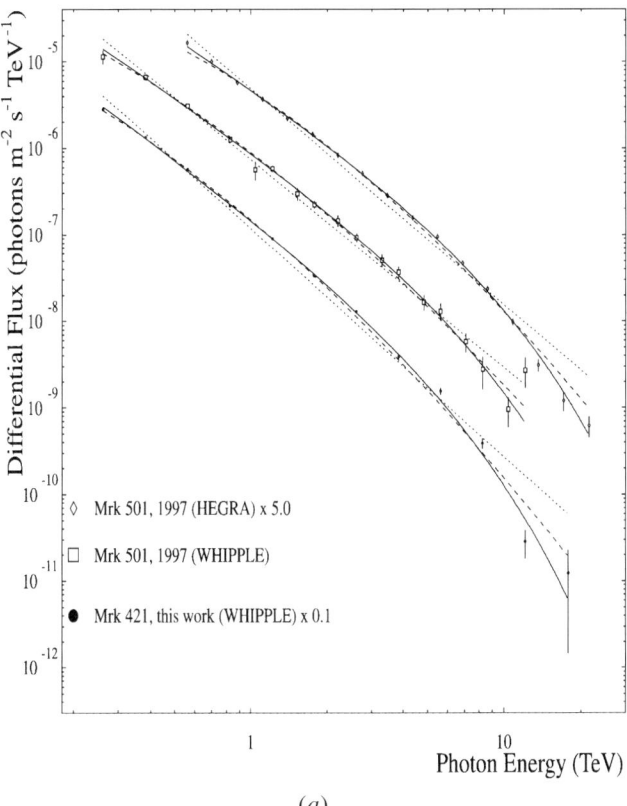

Figure 11.8. (*a*) Comparison of the spectra of Mrk501 and Mrk421 measured during the large outbursts in 1997 and 2001, respectively [1, 11]. For clarity, the spectra have been displaced vertically. All three measurements are best fit by a power law with exponential cut-off between 3 and 6 TeV (full curve). (*b*) Variations in the shape of the TeV energy spectrum as a function of total intensity; these data were recorded during the large outburst of Mrk421 in 2001 as seen at the Whipple Observatory [12]. (Reproduced with permission from the *Astrophysical Journal*.)

Attenuation of the VHE gamma rays by pair production with background infrared photons could produce a cut-off that is approximately exponential (chapter 14). The TeV signal recorded during some extraordinary flaring of Mrk421 in 2002 was sufficiently strong that the data could be divided and spectrally analysed; the results show that the spectrum clearly hardens with total intensity but the same exponential cut-off can be fitted to all the data [11].

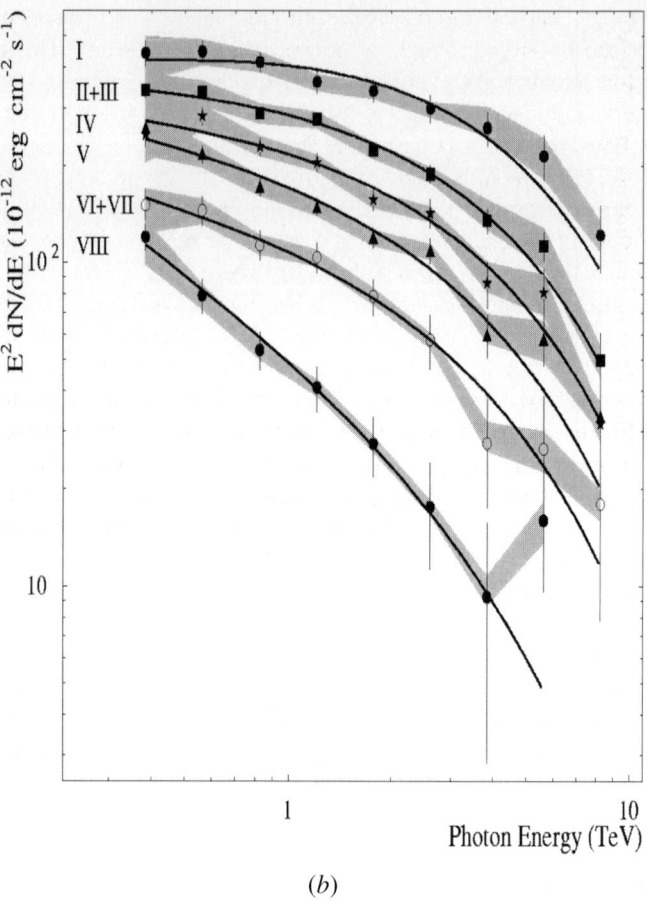

Figure 11.8. (Continued.)

11.3.7 Multi-wavelength observations

The problems associated with multi-wavelength campaigns organized to observe blazars across the spectrum are compounded in observing the VHE blazars because the sensitivity is such that time variations are seen on very short time scales. It is not clear if similar time variations are present at GeV energies but are not observed because of the relative insensitivity of MeV–GeV telescopes on space platforms. Nonetheless, because of the relative flexibility of observations with ground-based telescopes, it has been possible to organize some extensive campaigns so that the SEDs of the VHE blazars are much better determined than those of the HE blazars. It is also possible to study variations across a broad spectrum on shorter times scales.

One of the earliest multi-wavelength campaigns was organized in 1995 to measure the multi-wavelength properties of Mrk421 better. This campaign revealed, for the first time, correlations between VHE gamma rays and x-rays. Observations were conducted over a two-week period with the Whipple telescope, EGRET, ASCA (x-rays), the Extreme Ultraviolet Explorer (EUVE), an optical telescope, and an optical polarimeter. Observations with EGRET did not result in a detection of Mrk421. The 2σ flux upper limit for $E > 100$ MeV is 1.2×10^{-7} cm^{-2} s^{-1}, somewhat below the level detected in 1994. The light curves for some of these observations are shown in figure 11.9. Fortuitously, Mrk421 underwent a large amplitude flare in VHE gamma rays during the observation period. The flare is also clearly seen in the ASCA and EUVE observations. The x-rays and VHE gamma rays appear to vary together, limited by the one-day resolution of the VHE observations. The amplitude of the flaring is similar, with $\sim$400% difference between the peak flux and that at the end of the observations.

It is clear that this campaign undersampled the VHE part of the spectrum. This was remedied in a campaign, conducted in late April 1998, which again happened to coincide with a flare. Observations, at TeV energies with the Whipple telescope and at x-ray wavelengths with the BeppoSAX satellite, established the first hour-scale correlations between x-rays and gamma rays in a blazar. The light curve for the observations by BeppoSAX in three x-ray bands and Whipple above 2 TeV is shown in figure 11.10. The flare is clearly detected in x-rays and TeV gamma rays during the first day of observations. The peaks in the light curves occur at the same time, within 1 hr, but the fall off in the x-ray flux is considerably slower than the TeV gamma rays. Subsequent observations show that while there are strong correlations at x-ray and gamma-ray wavelengths, the correlations are complex.

Multi-wavelength observations of Mrk501 during its high emission state in 1997 revealed, for the first time, clear correlations between its VHE gamma ray and hard x-ray emission (figure 11.11). This time the observations were in the VHE range with the Whipple telescope and in the hard x-ray range with the OSSE telescope on the CGRO. Figure 11.11 shows daily flux levels for the contemporaneous observations of Mrk501. An 11-day rise and fall in flux is evident in the VHE and x-ray wavebands. The 50–150 keV flux detected by OSSE also increases. The optical data may show a correlated rise but the variation is small (at most 6%).

11.3.8 Spectral energy distributions

Figure 11.12 shows the SEDs expressed as power per logarithmic bandwidth, for Mrk421 and Mrk501 derived from contemporaneous multi-wavelength observations and an average of non-contemporaneous archival measurements. Both have a peak in the synchrotron emission at x-ray frequencies, which is typical of XBLs, and a high energy peak whose exact location is unknown but must lie in the 10–250 GeV range. Both the synchrotron and high energy peak are

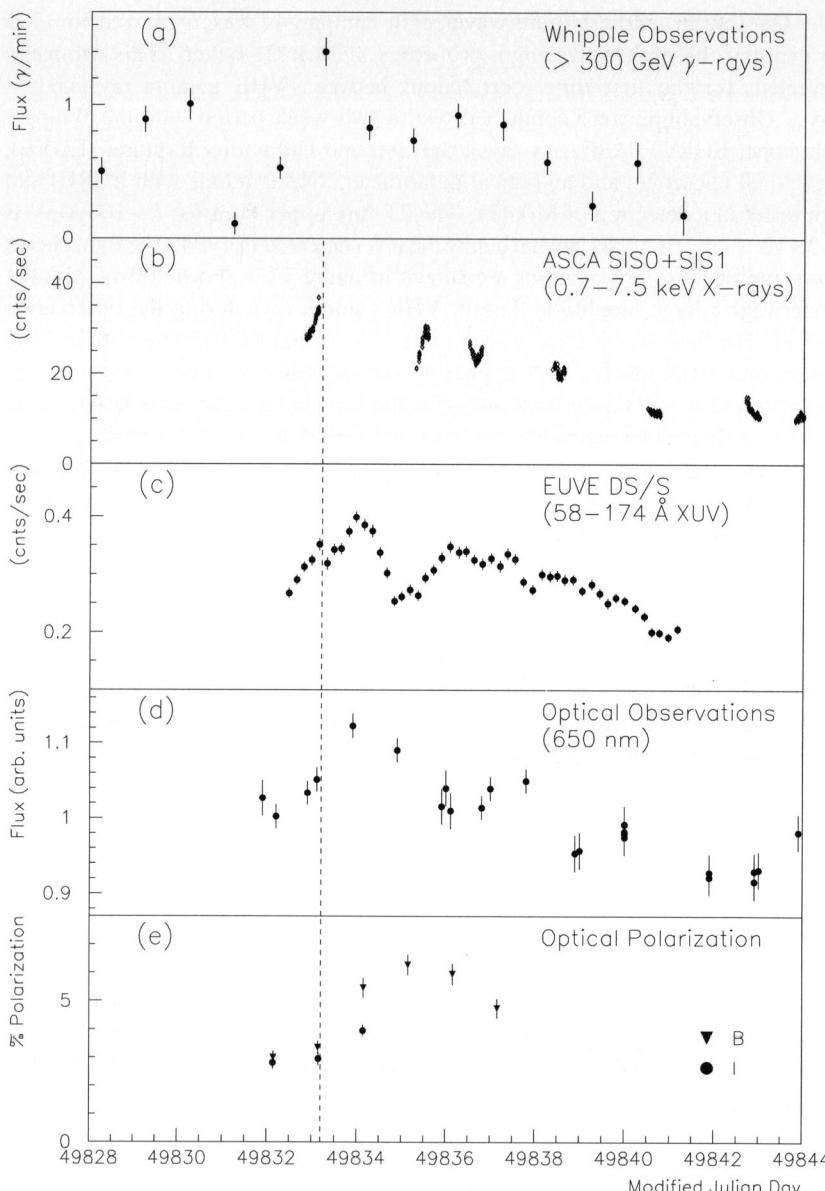

Figure 11.9. Multi-wavelength observations of a flare in Mrk421. (a) Gamma ray, (b) x-ray, (c) extreme-UV, (d) optical, and (e) optical polarization measurements of Mrk421 taken April–May 1995. 26 April corresponds to MJD49833 [2].

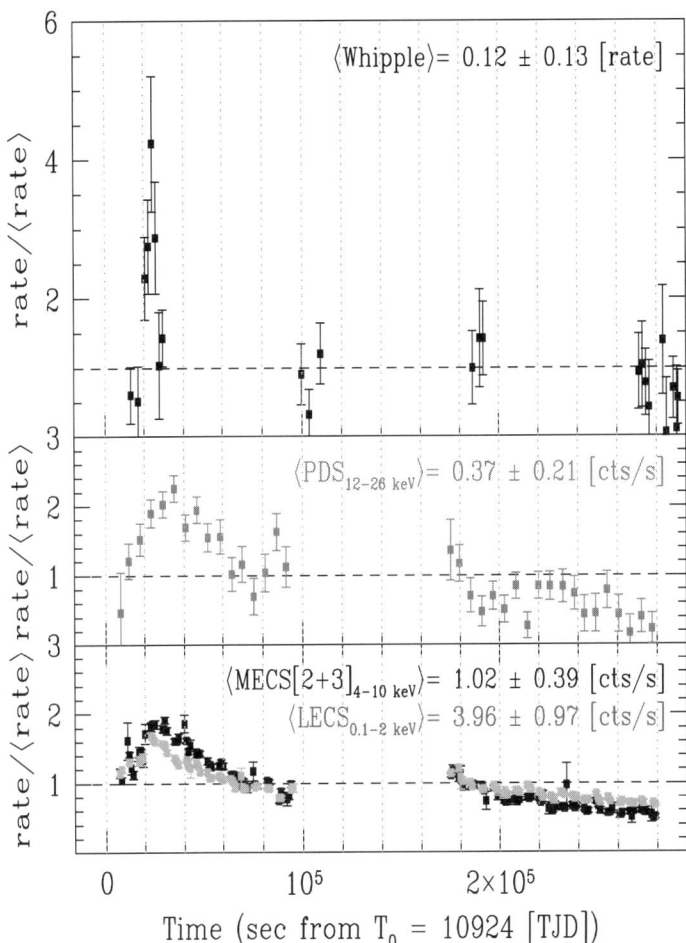

Figure 11.10. Light curves for observations of Mrk421 in 1998 April by Whipple and BeppoSAX. Whipple observations are for $E > 2$ TeV and are binned in 28 min observing segments. All count rates are normalized to their respective averages (listed at the top of each panel) for the observations shown [15]. (Reprinted from Catanese and Weekes 1999 *Publ. Astron. Soc. Pac.* **111** 1193. Copyright 1999 Astronomical Society of the Pacific; reproduced with permission of the editors.)

similar in power output, unlike the EGRET-detected FRSQs which can have high energy peaks well above the synchrotron peaks. Also, during flaring episodes, the x-ray spectrum in both objects tends to harden significantly, while the VHE spectrum is not observed to change.

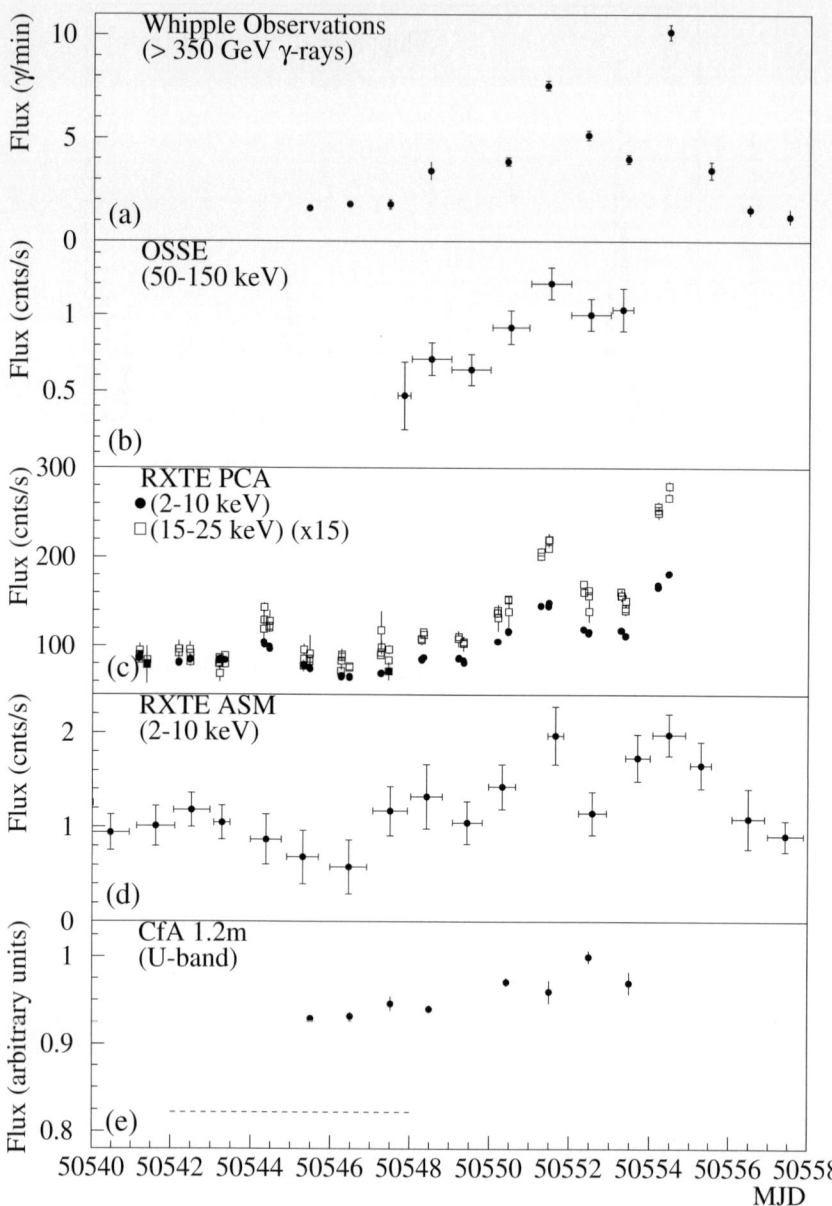

Figure 11.11. (a) VHE γ-ray, (b) OSSE 50–150 keV, (c) RXTE 2–10 keV and 15–25 keV, (d) RXTE All-Sky Monitor 2–10 keV and (e) U-band optical light curves of Mrk501 for the period 2 April 1997 (MJD 50540) to 20 April (MJD 50558). The dashed line in (e) indicates the average U-band flux in March 1997.

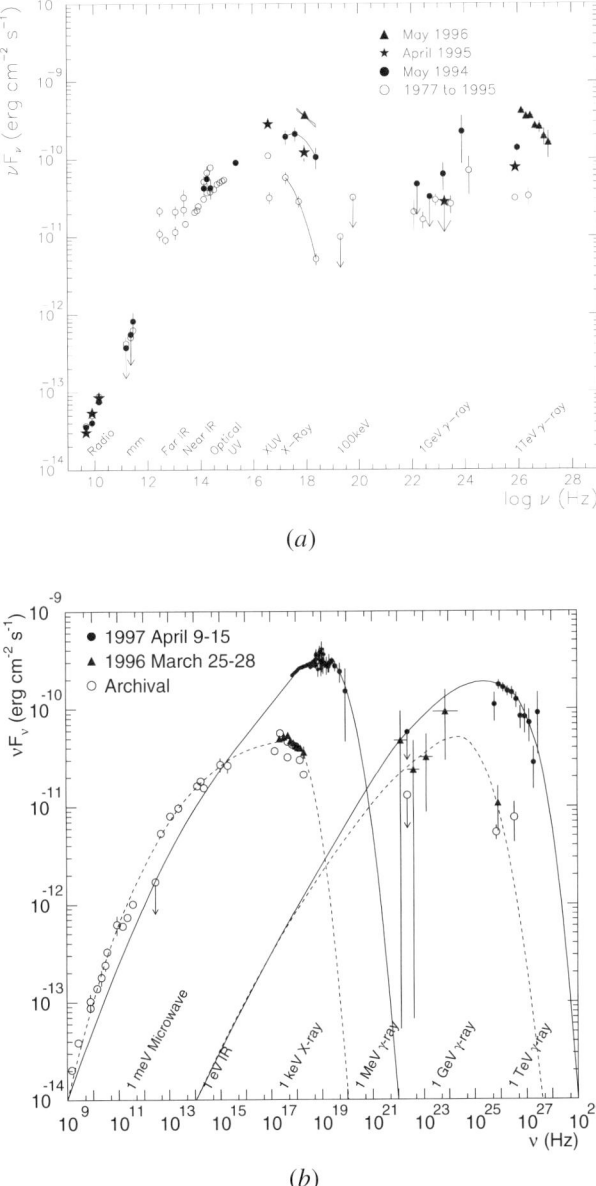

Figure 11.12. (a) The spectral energy distribution of Mrk421 from contemporaneous and archival observations [2]. (b) The spectral energy distribution of Mrk501 from contemporaneous and archival observations [3]. (Reprinted from Catanese and Weekes 1999 *Publ. Astron. Soc. Pac.* **111** 1193. Copyright 1999 Astronomical Society of the Pacific; reproduced with permission of the editors.)

11.3.9 Future prospects

With the new generation of ground-based arrays of telescopes now coming online (chapter 2), it is expected that the number of AGN detected above 100 GeV will increase tenfold. It is unlikely that AGN will be detected at these energies at distances beyond $z = 0.5$ so it will not be possible to study long-term evolutionary effects. However, the detection of short time variations out to these distances is potentially important for limiting models of quantum gravity. It may be possible to detect the weak signals expected from misaligned blazars. Spectral measurements will permit detailed models to be confronted with observations. The expected spectral cut-off between 10 and 100 GeV due to the extragalactic infrared background (chapter 14) will be measured.

Historical note: discovery of 3C279

Blazars made a dramatic entrance into the gamma-ray astronomer's lexicon within a few months of the launch of the CGRO, almost by accident. While the EGRET instrument was still being characterized, a target of opportunity was declared for observations of a new supernova in a relatively nearby galaxy [9]. The supernova, SN1991T, was not seen by EGRET but, as 3C273 (previously reported by COS B) was in the field of view, the data were immediately analyzed to confirm the earlier detection. To the experimenters' surprise, the preliminary analysis did not show 3C273 ($z = 0.158$) but did indicate the detection of another, more distant, blazar, 3C279 ($z = 0.538$). In the course of the two-week observation, the gamma-ray flux varied by a factor of five and the spectrum was seen to be very hard, much harder than 3C273. More intense data analysis showed that, in this one exposure, not only was 3C273 seen (but in a fainter state than the discovery observation ten years previous), but another, much more distant blazar, PKS0528+134 ($z = 2.06$) was also apparent. In this single exposure, much of the subsequent HE gamma-ray blazar phenomenon was elucidated: that extragalactic gamma-ray sources are detectable, that they can be seen out to $z>2$, that they are variable on a variety of times scales, and that gamma-ray blazars are plentiful.

References

[1] Aharonian F A *et al* 1999 *Astron. Astrophys.* **349** 11
[2] Buckley J H *et al* 1996 *Astrophys. J. Lett.* **472** L9
[3] Catanese M 1999 *BL Lacertae Phenomenon (ASP Conf. Series 159)* ed L O Takalo and A Silanpaa (San Francisco, CA: ASP) p 243
[4] Catanese M and Weekes T C 1999 *Pub. Astron. Soc. Pac.* **111** 1193
[5] Collmar W *et al* 2000 *5th Compton Symposium (AIP Proc. 510)* ed M L McConnell and J M Ryan (New York: AIP) p 303
[6] Gaidos J A *et al* 1996 *Nature* **383** 319

[7] Hartman R C *et al* 1999 *Astrophys. J. Suppl.* **123** 79
[8] Kataoka J *et al* 1999 *Astrophys. J.* **514** 138
[9] Kniffen D A *et al* 1993 *Astrophys. J.* **411** 133
[10] Kranich D *et al* 1999 *Proc. 26th ICRC (Salt Lake City)* **3** 358
[11] Krennrich F *et al* 2001 *Astrophys. J. Lett.* **560** L45
[12] Krennrich F *et al* 2002 *Astrophys. J. Lett.* **575** L9
[13] Lamb R C and Macomb D J 1997 *Astrophys. J.* **488** 872
[14] Fruin J H *et al* 1964 *Phys. Lett.* **2** 176
[15] Maraschi L *et al* 1999 *TeV Astrophysics of Extragalactic Sources (Astropart. Phys. 11)* ed M Catanese and T C Weekes, p 189
[16] Punch M *et al* 1992 *Nature* **358** 477
[17] Quinn J *et al* 1996 *Astrophys. J. Lett.* **456** L83
[18] Thompson D J *et al* 1995 *Astrophys. J. Suppl.* **101** 259
[19] von Montigny C *et al* 1995 *Astrophys. J.* **440** 525
[20] Weekes T C 2000 *Symposium on High Energy Gamma-Ray Astronomy (Heidelberg, June) (AIP Conf. Proc. 558)* ed F A Aharonian and H J Volk (New York: AIP) p 15

Chapter 12

Active galactic nuclei: models

12.1 Phenomenon

By any definition, AGN are extraordinary objects. Their extremely high luminosity, their spectacular images, their rapid and random time variability, their energy spectra, which belie thermal processes and indicate the presence of gigantic particle accelerators, mark them as some of the most important denizens of the cosmic zoo. The presence of a supermassive black hole, an accretion disk, a dusty torus, emission line clouds, a relativistic jet—these all indicate extraordinary complexity and a wonderful laboratory for astrophysical research. Before the advent of gamma-ray astronomy as a truly observational science, the basic phenomenon of AGN was known. It was the optical study of flat-spectrum radio sources (FSRS) that first drew attention to AGN since it was seen that the intense non-thermal radiation outshone the normal stellar population by many orders of magnitude. The observation of hot accretion disks, broad emission line clouds, and polarized radiation from the radio bands through x-rays heightened the interest.

The discovery that AGN were luminous in gamma rays over many decades, up to the highest energies, has opened a new window for the investigation of these objects; it confirms the richness of the phenomenon but also deepens its mystery. In the unified picture of AGN, it is probable that all AGN, at some level, are emitters of high energy gamma rays and that only some are detected is an accident of orientation, of geometry, not of physics. As with pulsars (see chapter 8), if AGN had been first detected as gamma-ray sources, it is probable that they would have been regarded as primarily 'gamma-ray sources' with the other radiations somewhat auxiliary, since the gamma rays indicate the highest energy activity and, therefore, the greatest challenge to the energy acceleration machine.

Blazars, the gamma-ray observable members of the AGN family, are unique in that they can be observed over the full range of the electromagnetic spectrum, some 19 decades of energy. The bulk of the observed radiation is non-thermal, which indicates that we are dealing with a relativistic phenomenon and that, given the energetics, an important extragalactic source of cosmic radiation. The gamma-ray luminosity, L_γ is very large. Assuming isotropic emission, L_γ can be as high as 10^{49} erg s^{-1}. Even if the emission is beamed with a beaming factor of 10^{-3}, then $L_\gamma \approx 10^{46}$ erg s^{-1}, still a prodigious amount of radiation, particularly since it implies even greater total power in the relativistic particles that produce this exotic radiation.

A feature of the blazars is their variability; this is observed at all wavelengths in which there are sensitive detectors. The observed time scales of variability span many decades and there are complex, but strong, correlations across many bands. An analysis of the characteristic time variations in these bands does not indicate strong evidence for periodic variations but it does appear that, in many bands, there is a break in the time power spectrum near 10^5 s. The physical implications of this time are not understood but it may indicate a preferred length scale of 1 light-day.

12.2 Source of energy

Although the supermassive black hole, that is at the heart of the AGN, is the putative source of energy for the entire system, it is not intuitively obvious how energy can be extracted from the black hole [1]. In particular, it is not obvious why it should emerge in the form of a relativistic jet. Black holes are generally conceived as energy sinks—a gravitational sink hole that pulls in everything in its vicinity. While the formation of a hot accretion disk around the black hole is understandable in terms of conservation of angular momentum and, hence, the reason that these objects are detectable as bright x-ray sources, it is less clear why beams of relativistic particles should *emerge* from the vicinity of the black hole.

The formation of jets is now regarded as an ubiquitous phenomenon. They are seen on a much smaller scale in stellar systems where the outflow, although supersonic, is still much less than that of the velocity of light (chapter 9). Nonetheless, it is believed that similar mechanisms are at work on both the stellar and galactic scale. In fact, it is the study of nearby stellar systems such as SS433, in which rotating beams of high energy particles (whose velocities reach $0.25c$) are clearly seen, that has elucidated the overall phenomenon.

Well before the identification of blazars, it had been established that the large radio lobes associated with relatively nearby radio galaxies were formed by the collisions of beams of fast moving particles with the swept-up intergalactic gas. The particles were assumed to be emitted by some mechanism in the core of the galaxy. The source of the energy that propelled these beams, whose constituents moved at supersonic velocities and close to the speed of light, was not clear. When it became clear that the ultimate engine in these systems was a supermassive black hole and that the beams originated close to the black hole, it was assumed that the formation of the jets was beyond the realm of observers and that subsequent research would be on the basis of theoretical investigations. This is still largely true, although observations now probe remarkably close to the base of the beams. Radio techniques can discern angular distances scales of as small as 10^{-5} arc-sec. Although the event horizon of the black hole in the nearby radio galaxy, Cygnus A, is still much smaller than that (a few light-hours), it has become possible to see amazing detail in the inner workings of AGN. With Doppler beaming effects at play in BL Lac objects, we can effectively 'see' even closer with gamma-ray observations [1].

The challenges for understanding the relativistic jets that are the heart of blazars are: what is the source of the energy; how can these relativistic energies be achieved; what is the reason that two jets are formed pointing in exactly opposite directions; and what is the mechanism by which the jets remain collimated as they emerge through the turbulent surroundings of the black hole. The two opposite jets are probably aligned along the opposite poles of rotation of the black hole. The explanation of the extended jet collimation is difficult but the energy source of the beams can be understood to originate from one of the few other properties that black holes possess in addition to their mass: rotational energy.

The rotational energy can arise from the residual angular momentum either as the black hole formed out of an accreting cloud of gas, or as a result of the accreting extragalactic material, or even because of a merger of two or more black holes.

Models that have been proposed include hydrodynamic models involving thermal pressure in the accreting gas and mechanisms that are specific to black holes, arising from the impossibility of a black hole accreting magnetic flux and having to eject the magnetic flux carried in by the accreting gas. Although the exact details of the models have not been agreed upon, the concepts has been successful in providing some explanation for the formation of jets on a variety of scales from stellar systems such as proto-stars, that do not contain black holes at their cores, to AGN, with supermassive black holes at their centers.

12.3 Beaming

Though there is no general consensus on the origin of the emission components seen in gamma-ray blazars, it is generally agreed that the low energy component arises from incoherent synchrotron emission by relativistic electrons within the jet. This is supported most strongly by the high-level variable polarization observed in these objects at radio and optical wavelengths. The observation of HE and VHE gamma rays from blazars is strong and independent evidence that the radiation in blazars is produced in relativistic jets and that the jets make a small angle with the line of sight. This conclusion is also drawn from the inferred compactness of the source emission region based on the observed short-term variability (light-minutes to light-hours); given the observed gamma-ray luminosity (10^{48}–10^{49} erg s^{-1}), it would not be possible for the gamma rays to emerge from the source without absorption by gamma–gamma pair production unless there is relativistic beaming.

It is generally agreed that the relativistic jets are caused by bulk relativistic motion which is characterized by the bulk Lorenz factor, Γ, defined by

$$\Gamma = (1 - \beta^2)^{-1/2}.$$

The Doppler factor, δ, of an object moving at $\beta = v/c$, making an angle θ with the line of sight, is then

$$\delta = 1/(\Gamma(1 - \beta \cos \theta)).$$

Hence, if $\beta = 0.95$ and $\theta = 5°$, i.e. viewed almost directly down the jet, $\Gamma = 3.2$ and $\delta = 5.8$. The introduction of a Doppler factor alleviates the explanation of the observed properties of blazars in a variety of ways [4]. The very short time variations, dt_{obs} seen at high energies introduce a problem in that the emission appears to come from a small volume with dimension $D = dt_{obs}c$; the density of low energy synchrotron photons may be so great that the gamma rays must pair produce and cannot escape. But, with relativistic boosting, $dt_{obs} = t_{source}/\delta$, and it is possible to avoid the so-called Compton catastrophe.

Relativistic boosting effectively removes several other embarrassing properties of the observations. The problem of accelerating particles to such high relativistic energies is alleviated by the fact that the observed frequency v_{obs} equals δv_{source}, i.e. the emitted photon is a factor of $1/\delta$ less than that actually emitted in the rest frame. Hence, the maximum energies that must be achieved are reduced.

The observed luminosity is even more dramatically reduced since from a combination of factors, the observed luminosity L_{obs} equals $\delta^4 L_{source}$.

In practice, often $\delta \approx 10$–20, so that the enhancement in energy can be as much as 10^{3-4}. One can think of this as a way of amplifying the normally faint regions close to the central black hole and thus permitting a view of conditions that would otherwise be unobservable. This is not unlike the way in which gravitational lensing permits a view of otherwise unobservable distant faint sources in the universe. In both phenomena special and general relativity are at work [2].

Superluminal motion (see historical note: superluminal motion) can be used to get an independent measure of Γ. It is difficult to observe superluminal motion in blazars since they are viewed at $\theta \approx 0$. In the GeV blazar, 3C273, the measured value of β' is 10, indicating that $\beta = 0.995$; in 3C279 $\beta' = 4$. Radio observations of Mrk501 give $\beta' = 6.7$.

12.4 Models

The origin of the high energy emission is one of the most challenging features of the gamma-ray-emitting blazars in that it is the most difficult to explain and,

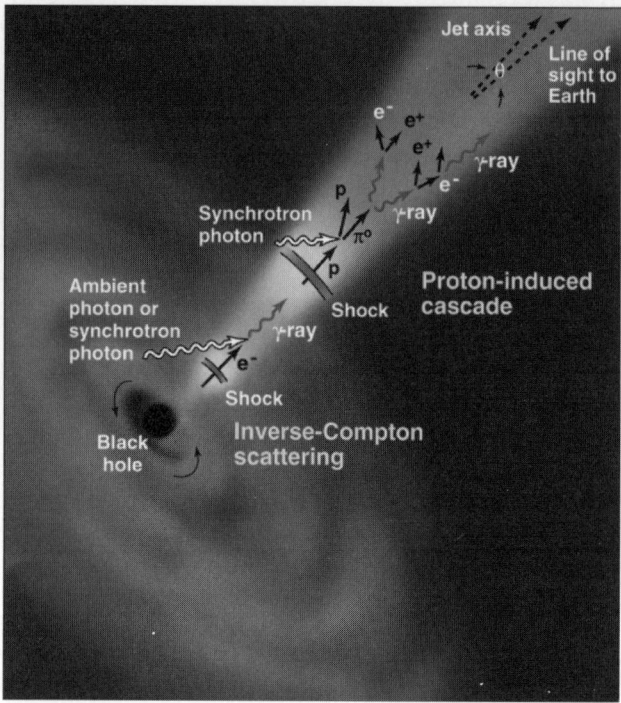

Figure 12.1. Gamma-ray production in relativistic jets. (Reprinted from Buckley J 1998 *Science* **279** 676. Copyright 1998 American Association for the Advancement of Science.)

at the same time, intrinsic to our understanding of these objects. There are many variations to the models that have been proposed and here mention is made only of the main features of models which are most often invoked to explain the gamma-ray emission. The general phenomenon of gamma-ray production in relativistic jets is illustrated in the cartoon in figure 12.1 [6].

12.4.1 Lepton models

The characteristic shape of the SED of GeV and TeV blazars gives an immediate hint as to the probable radiation mechanisms at work. The double-peaked spectrum is immediately suggestive of the Compton-synchrotron model that appears to work so well in modelling plerions like the Crab Nebula (figure 12.2). Although the conditions in the relatively stable environment of a plerion are quite unlike those in the chaotic conditions of a relativistic jet, the basic physical mechanisms appear to be the same. Electrons are accelerated beyond the velocity of the bulk Lorenz outflow. This acceleration is probably produced by shocks propagating down the jet which are caused by colliding inhomogeneities in the jet,

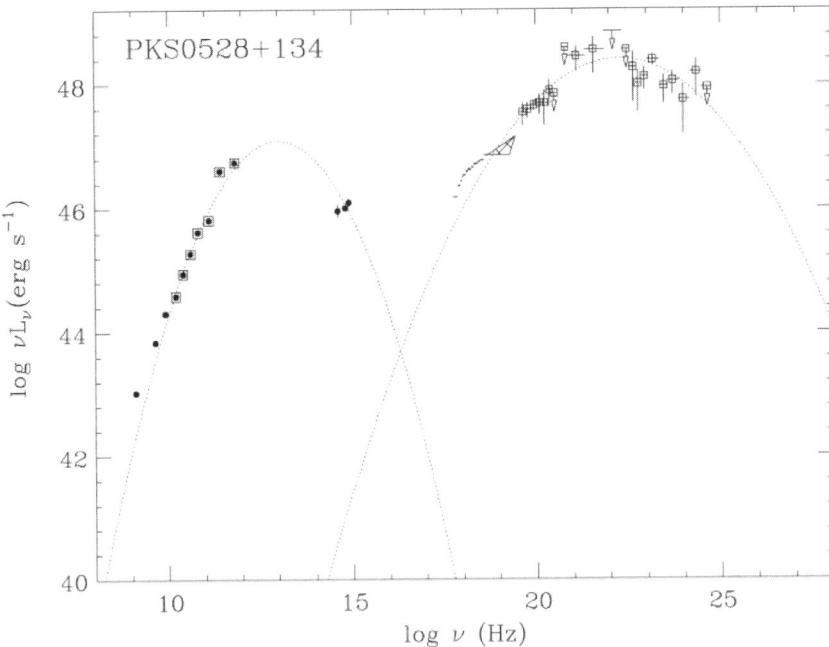

Figure 12.2. The typical SED of a blazar detected by EGRET, in this case PKS0528+134. The luminosity is clearly greatest at GeV gamma-ray energies (Compton peak) with the lesser synchrotron peak in the infrared-microwave region [10]. (Reproduced with permission from the *Astrophysical Journal*.)

i.e. blobs of material moving down the jet with different velocities. The electrons radiate synchrotron radiation in the magnetic fields associated with the jet and thus produce the first peak, ν_{synch}, in the SED. The observation of polarization is some confirmation that this mechanism is at work. The position of this synchrotron peak is determined by the efficiency of the shock acceleration processes and the cooling (energy loss) processes. The cooling comes from synchrotron energy loss and Compton scattering (see appendix).

12.4.1.1 Synchrotron self-Compton (SSC)

Compton scattering is inevitable since the synchrotron-radiated photons themselves will provide soft photon targets to be boosted to energies close to that of the radiating electrons with Lorentz factor, γ [13]. This energy is given by

$$E_\gamma \approx \gamma^2 h\nu$$

in the regions where Thompson scattering prevails and

$$E_\gamma \approx \gamma m_e c^2$$

in the Klein–Nishina region.

In the simplest 'one-component' model, it is assumed that the synchrotron radiation and Compton scattering take place in the same region; realistic models are more complex and require multiple shells. Before the detection of the GeV blazars by EGRET, it was assumed that almost all the observed high energy properties of blazars could be explained in terms of SSC models. However, the large luminosity of the GeV blazars could not be accounted for by SSC models and other processes had to be invoked.

The peak in the synchrotron spectrum is given by $\nu_{\text{synch}} = 2.8 \times 10^6 \gamma'^2 B \delta$ Hz for a sphere with uniform magnetic field, B, and a power-law electron spectrum which breaks at γ'. Although there is some dependence on δ and, hence, on the geometry of the source and the viewing angle, it is not sufficient to explain the wide range of observed ν_{synch}. This can only be explained by a change in the physical parameters, B and γ'.

12.4.1.2 External radiation Compton (ERC)

In the case of plerions, it was observed that the Compton-synchrotron models had often to take account of collisions of the relativistic electrons with soft photons from the microwave background which, of course, permeates all space. In some cases, these 'external' photons were denser than the synchrotron soft photons and were the dominant targets for Compton scattering. These have a negligible effect in AGN but other sources of external soft target photons in the AGN itself can play a role. Hence, the second class of models involve external radiation Compton [16, 8]. The existence of these sources of soft radiation is deduced from observations of non-blazar AGN i.e. those in which the jet is not seen edge-on. The SEDs of these sources show additional thermal bumps in the UV (probably the accretion disk), in the infrared (the hot torus) and in x-rays (the disk corona). The ERC models in which the gamma-ray emission arises predominantly from inverse Compton scattering of seed photons which are produced outside of the jet, directly from an accretion disk, after being re-processed in the broad-line region or scattering off thermal plasma, fit the GeV observations better than the SSC models.

12.4.2 Proton models

Another set of models proposes that the gamma rays are produced by proton-initiated cascades [11, 12]. These models are strongly motivated by the desire to explain two puzzling extragalactic phenomena simultaneously: the production of VHE gamma rays in AGN and the origin of the extragalactic cosmic radiation with energies up to 10^{20} eV and beyond. From energy considerations, AGN are the most likely source of the extragalactic cosmic radiation and it would be convenient if the VHE gamma-ray observations could be seen as a direct confirmation of the acceleration of hadrons in AGN. A variety of mechanisms have been proposed but none have found universal favor.

The production of VHE gamma rays can be modelled in terms of the acceleration of protons in the jet to energies of 10^{18} eV. The protons interact with soft photons in the jet and produce mesons by photoproduction.

$$p + h\nu \rightarrow p + \pi^0 \text{ or } n + \pi^+.$$

The pions produce cascades which can, in principle, explain many of the features in the observed SEDs. The lower energy radiation is produced by synchrotron radiation from the secondary products of the cascade. In some models, the gamma rays are synchrotron radiation from the protons whose energies can reach up to 10^{20} eV. While these proton models have no problem explaining the highest gamma-ray energies, they do have problems accounting for the rapid cooling necessary to account for the short time variations observed. Proton cooling by synchrotron radiation is less than that by electrons by the factor $(m_e/m_p)^3$. However, cooling can also come from collisions with photons or ions.

A by-product of proton models is that the decay of charged pions will produce energetic neutrinos which might be detectable with the next generation of neutrino telescopes. The detection of such a flux would effectively eliminate the lepton models. This possibility is the mainstay of theoretical predictions of extragalactic neutrino fluxes and is often quoted as the justification for the construction of large neutrino telescopes.

The HE and VHE gamma-ray observations strain both the lepton and proton models but do not, at present, rule either out. However, as we will see later, the observations generally seem to favor lepton models.

12.5 Implications of the gamma-ray observations

12.5.1 HE observations

The interpretation of the data from the many blazars detected by EGRET is complicated by the fact that the exposures are short (approximately two weeks) and the objects are generally detected in a flaring state without contemporaneous data at other wavelengths. It is difficult, therefore, to determine the full SED. The observed fluxes can vary by two decades. The shortest time variations observed (approximately a few hours) limit the emission regions to less than the dimensions of the supermassive black hole ($\approx 3 \times 10^{10}$ km if the black hole mass is $>10^9$ M$_\odot$). Doppler beaming reduces this limitation. The rapidity of the time variations is not as extreme as that seen in x-rays or VHE gamma rays but it is not clear whether this is a limitation of the sensitivity of the VHE detector or inherent to the sources.

Although in some cases extensive observing campaigns have been organized to monitor particular sources, the results have often been ambiguous. The major implication of the GeV observations have been the large luminosities observed and the large range of distances over which the phenomena are observed [17]. A study of the variation of the observed power-law spectral index with distance does not show any significant trends within the limited accuracy of the measurements and, hence, does not lead to any conclusions about the evolution of blazars out to $z = 2.5$. Because of the large populations of observed sources, it has been possible to do detailed correlations with the observed averaged properties at other wavelengths.

The observed SEDs indicate that these sources cannot be fitted by simple SSC models. Other sources of soft radiation are brighter than the synchrotron photon density and, hence, ERC models are favored. Although EGRET detected some 70 AGN, only for a few are there sufficient contemporaneous data to permit detailed model fits.

GeV gamma rays pair produce with keV photons. Since the density of the x-ray photons in these sources is not large, there is no strong limitation on the escape of the gamma rays and, hence, the total gamma-ray luminosity, L_γ, can be large.

PKS 0528+134 was observed in two extreme states: a high state in March 1993 and a low state in February 1997. There was a difference in L_γ between the two states of a factor of 50. A two-component model, which took into account a combination of SSC and ERC emission, gives a reasonable fit to the data [14]. In the high state, the ERC component dominates the emission but the SSC component is the dominant mechanism in the low state. In the low state, ν_{synch} moves to higher frequencies. A similar relationship is observed in 3C279, another of the relatively well-observed EGRET blazars.

12.5.2 VHE observations

The observations of VHE AGN, although limited in number and confined to relatively nearby objects, have already significantly limited the parameters of possible models of BL Lac objects. Although the small numbers prohibit population or evolution studies, the VHE observations have been more restrictive on source models than the HE observations [7]. For example, the rapid variability indicates either very low accretion rates and photon densities near the nucleus or, conversely, requires the gamma-ray emission region to be located relatively far from the nucleus to escape the photon fields. The VHE gamma-ray blazars are significantly less luminous in gamma rays than the HE blazars. In VHE AGN, L_γ is of the same order as x-ray luminosity, L_x, whereas in the HE blazars L_γ is the dominant component, often by a factor of ten.

The observation of VHE emission from blazars has helped resolve the nature of the differences between the LBLs and HBLs. Based on their smaller numbers and higher luminosities, it had been proposed that LBLs were the same as HBLs but with jets aligned more closely with our line of sight. However, the observation of rapid variability and spectra that extended up to TeV energies in HBL blazars point to the differences between the two sub-classes being more fundamental: the

HBLs have higher maximum electron energies and lower intrinsic luminosities.

Simultaneous measurements of the synchrotron and VHE gamma-ray spectra constrain the magnetic field strength, B, and Doppler factor, δ, of the jet. The correlation between the VHE gamma rays and optical/UV photons observed from Mrk421 indicates that if both sets of photons are produced in the same region of the jet, $\delta \gtrsim 5$ is required for the VHE photons to escape significant pair-production losses. If the SSC mechanism produces the VHE gamma rays, $\delta = 15$–40 and $B = 0.03$–0.09 G for Mrk421 and $\delta \approx 1.5$–20 and $B = 0.08$–0.2 G for Mrk501. To match the variability time scales of the correlated emission, proton models which utilize synchrotron cooling as the primary means for proton energy losses require magnetic fields of $B = 30$–90 G for $\delta \approx 10$ [5]. The Mrk421 values of δ and B are extreme for blazars but they are still within allowable ranges and are consistent with the extreme variability of Mrk421.

In addition, the VHE observations have constrained the types of models that are likely to produce the gamma-ray emission. For instance, the correlation of the x-ray and the VHE flares is consistent with SSC models where the same population of electrons radiate the x-rays and gamma rays. The relative absence of flaring at EGRET energies can be explained because the lower energy electrons, which produce the gamma rays in the EGRET range, radiate away their energy more slowly than the higher energy electrons which produce the VHE emission. The MeV–GeV emission could be the superposition of many flare events and would, therefore, show little or no short-term variation.

In the ERC models, in which gamma rays are produced through the Compton scattering of external photons, the target photons must have energies <0.1 eV (i.e. in the infrared band) to avoid significant attenuation of the VHE gamma rays by pair production. There is little direct observational evidence of such an infrared component in BL Lac objects but the existence of such a field has been predicted as a product of accretion in AGN.

Models, in which the gamma-ray emission is produced by proton progenitors through e^+e^- cascades originating close to the base of the AGN jet, have difficulty explaining the TeV emission observed in Mrk421. The high densities of unbeamed photons near the nucleus, such as those from the accretion disk or the broad emission line clouds, cause high pair opacities to TeV gamma rays. These models predict that the radius at which the optical depth for gamma–gamma pair production drops below unity increases with increasing gamma-ray energy and, therefore, the VHE gamma rays should vary either later or more slowly than the MeV–GeV gamma rays [3]. This is in contradiction to the observations of Mrk421.

Although Mrk421 and Mrk501 are often considered to be very similar, in fact the gamma-ray observations have shown them to be quite different. Mrk421 has shorter time variations than Mrk501 and ν_{synch} is always in the soft x-ray band. The value of ν_{synch} in Mrk421 does not show the dramatic shifts seen in Mrk501 (from <3 keV to >50 keV) and the mean flux from Mrk501 is always less than that of Mrk421.

Table 12.1. LBL and HBL blazars.

Property	LBL	HBL
Gamma-ray emission	GeV	TeV
Initial selection	Radio	X-ray
Designation	Red	Blue
ν_{synch}	low	high
ν_{IC}	low	high
L_0 (broad emission lines)	high	none
L_γ	high	low
L_{IC}/L_{synch}	high	low

12.5.3 Unified theories

Blazars are difficult to detect and the early surveys (generally radio, optical, or x-ray) had led to the conclusion that there were two distinct classes of blazar which were differentiated by whether their radiation peaked at low (LBL) or high (HBL) frequencies. In practice, this corresponded to whether the blazar was discovered in a radio or x-ray survey. These sub-divisions have also been called 'red' and 'blue'. Their properties are summarized in table 12.1.

Historical note: superluminal motion

The phenomenon of superluminal motion was predicted before it was observed in a seminal paper by Martin Rees [15]. While not yet observed in gamma rays, it is seen at radio and optical wavelengths in many blazars and has direct relevance to the understanding of the properties of the jets. Radiating material in jets is observed to move relative to their cores at velocities that appears to exceed the velocity of light. This was first observed in VLBI studies of the blobs of radio-emitting material from radio sources whose jets were observed almost edge-on. It has since been seen in many optical jet sources, e.g. M87. It has a ready explanation in terms of the geometry of the moving material and involves no new physics. The geometry of the phenomenon is illustrated in figure 12.4. The apparent superluminal motion is characterized by β' defined by

$$\beta' = (\beta \sin\theta)/(1 - \beta \cos\theta) < \Gamma\beta.$$

Hence, by observing β', a lower limit to Γ can be determined. If $\beta = 0.95$ and $\theta = 5°$, $\Gamma = 3.2$, and $\beta' = 1.54$, i.e. the velocity of light is apparently exceeded by a factor of 1.54.

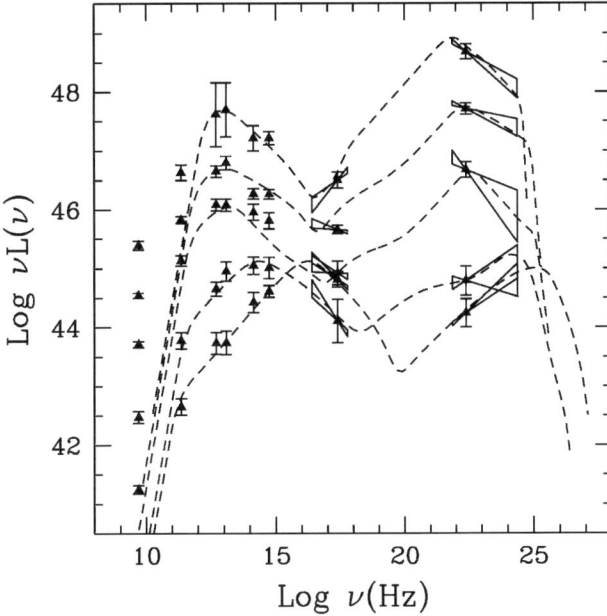

Figure 12.3. The SEDs of several blazars showing the gradual progression of the synchrotron and Compton peaks to increasing energy as the overall flux decreases [9]. (Reproduced with permission from the *MNRAS*.)

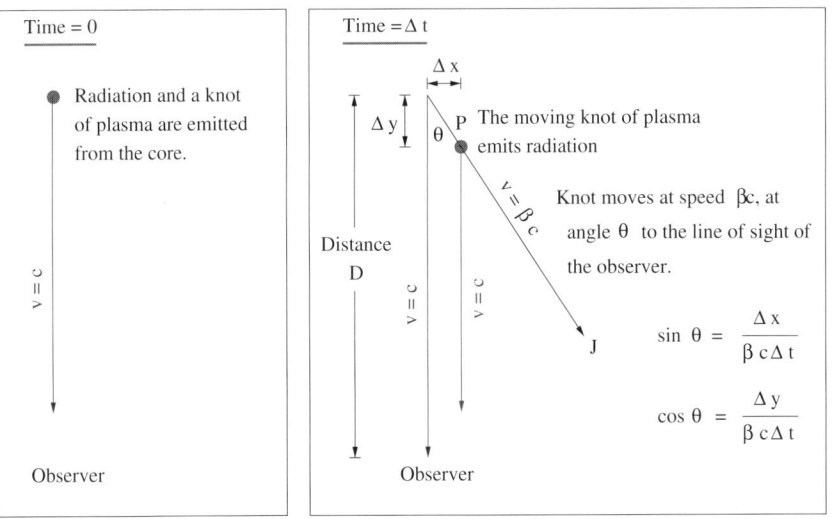

Figure 12.4. Geometry of the superluminal motion phenomenon. (Figure: D Horan.)

This differentiation is now believed to be artificial and caused by a selection effect. It appears that an objective survey would find that there was a continuum of properties ranging between the two SEDs shown in figure 12.3. It is significant that all the GeV blazars have been classified as LBL and the bulk of the TeV blazars as HBL. Thus there are not two distinct classes of BL Lacs but a continuum of objects [9]. The apparent sub-division is a selection effect which is disappearing as the surveys become less band-dependent and more complete.

What is clearly apparent is that there is a continuum of peak frequencies which scales with a continuum of overall power. The AGN with lower luminosity accelerate electrons to higher individual energies (which produce the VHE gamma rays) whereas the more luminous AGN are less efficient accelerators of high energy electrons or are more heavily absorbed and are detectable primarily as HE gamma-ray emitters.

References

[1] Begelman M and Rees M 1995 *Gravity's Fatal Attraction* (Scientific American Library)
[2] Bicknell G V, Wagner S J and Groves B 2000 *High Energy Gamma-Ray Astronomy (AIP Conf. Proc. 558)* ed F A Aharonian and H J Volk (New York: AIP) p 261
[3] Blandford R D and Levinson A 1995 *Astrophys. J.* **441** 79
[4] Blandford R D and Rees M J 1978 *Pittsburgh Conf. on BL Lac Objects* ed A M Wolfe (Pittsburgh, PA: University of Pittsburgh Press) p 328
[5] Buckley J *et al* 1997 *Proc. 4th Compton Symposium (AIP Conf. Proc. 410)* ed C D Dermer, M S Strickman and J D Kurfess (New York: AIP) p 1381
[6] Buckley J 1998 *Science* **279** 676
[7] Catanese M and Weekes T C 1999 *Publ. Astron. Soc. Pac.* **111** 1193
[8] Dermer C D, Schlickeiser R and Mastidchadias A 1992 *Astron. Astrophys.* **256** L27
[9] Ghisellini G *et al* 1998 *Mon. Not. R. Astron. Soc.* **301** 451
[10] Kubo H *et al* 1998 *Astrophys. J.* **504** 693
[11] Mannheim K 1993 *Astron. Astrophys.* **269** 67
[12] Mannheim K 1998 *Science* **279** 684
[13] Maraschi L, Ghisellini G and Celotti A 1992 *Astrophys. J. Lett.* **397** L5
[14] Mukherjee R *et al* 2001 *Proc. Gamma Ray Astrophysics 2001 (Baltimore, MD, April) (AIP Conf. Proc. 587)* ed S Ritz, N Gehrels and C R Schroder (New York: AIP) p 304
[15] Rees M J 1966 *Nature* **211** 468
[16] Sikora M, Begelman M C and Rees M J 1994 *Astrophys. J.* **421** 153
[17] von Montigny C *et al* 1995 *Astrophys. J.* **440** 525

Chapter 13

Gamma-ray bursts

13.1 Introduction

The study of gamma rays from astronomical sources has made many contributions to astronomy but none has been so startling and so revolutionary as the study of gamma-ray bursts (GRBs). The discovery of GRBs has provided one of the great puzzles in astrophysics in the last few decades and it has introduced the astronomical community to a whole new concept in astronomy: the astronomy of objects seen only once and then only for a very short time (figure 13.1). That these fleeting bursts should not be a local phenomenon but should originate in sources at cosmic distances is even more surprising. They are the most luminous emissions in the universe in any wavelength band and are perhaps the brightest phenomenon since the Big Bang. Because they are at cosmological distances, they offer a new tool for the exploration of objects on the edge of the observable universe.

Although the GRB phenomenon is usually associated with energies of 50 keV to 1 MeV (hard x-rays to low energy gamma rays), results from the Solar Maximum Mission (SMM) and from the EGRET detector on the Compton Gamma-Ray Observatory show that there is a component at high energies and thus the phenomenon must be included in this text on very high energy gamma rays. The power spectrum certainly peaks in the lower energy ranges (figure 13.2) but the observations at high energies really tax the models and may ultimately expose the underlying emission mechanism. Before discussing the high energy observations, an account will be given of the discovery of GRBs, their short but controversial history on the frontiers of astrophysical research, and the elucidation of their origins from the study of counterparts at other wavelengths.

13.2 The discovery

The discovery of GRBs ranks as one of the classic tales of scientific discovery in the 20th century. Like many of the most exciting discoveries, it was serendipitous.

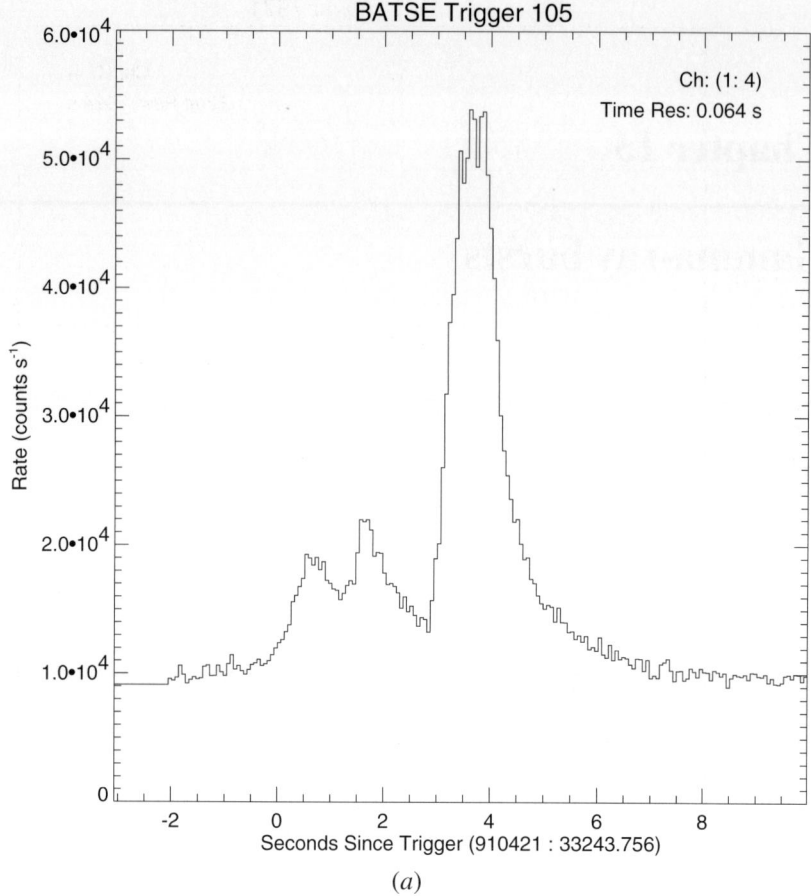

Figure 13.1. Light curves of four typical bursts recorded by BATSE. (Figure: BATSE/NASA.)

In fact, it was a happy outcome from a rather sad period in human history. In the early 1960s, the Cold War was at its peak and suspicion and distrust were evident on all sides. One bright spot was the signing of the nuclear test ban treaty in which each side undertook to make no more tests of nuclear weapons in the atmosphere or in space. A noble objective but how could one trust the other side when ideologies were completely opposed and memories of the Second World War were still fresh? The US Defense Department solution was to set up a monitoring system that would detect any breaches of the treaty by deploying a variety of sensors on spacecraft that would be sensitive to all such explosions in the near space environment. Space science was still in its infancy and the satellites were very small by today's standards. Nonetheless, the surveillance satellites

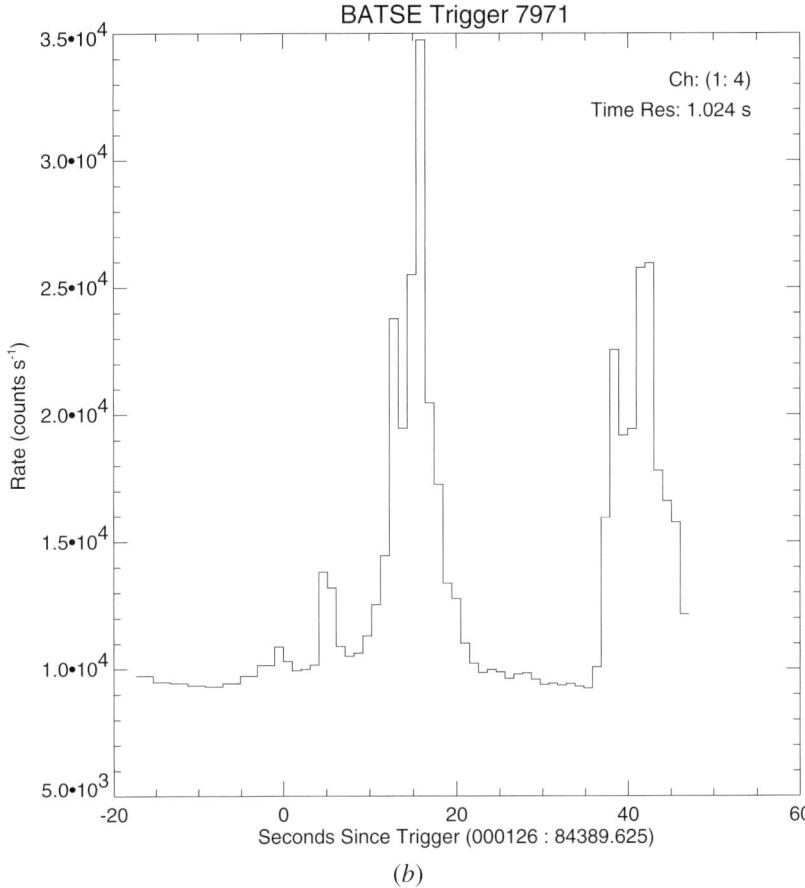

Figure 13.1. (Continued.)

(which were called Vela from the Spanish word, *velar*, to watch) were equipped with a variety of detectors that would detect the instantaneous radiation from the nuclear blast at x-ray, gamma ray, and optical wavelengths. By having two identical satellites on opposite sides of the earth, in circular orbits of 250 000 km in diameter, there was almost complete coverage of the earth and near space.

However, a clever enemy would know that the nuclear blast might be shielded locally or the test might be conducted behind the moon. In that case the prompt nuclear blast of x-rays would not be seen directly by the Vela satellites. Inevitably, a cloud of radioactive material would also be produced in the blast and this would be almost impossible to conceal. The delayed gamma radiation from radioactive decay could be detected if the satellite was suitably equipped to identify it. Fortunately for gamma-ray astronomy, the Vela series of detectors

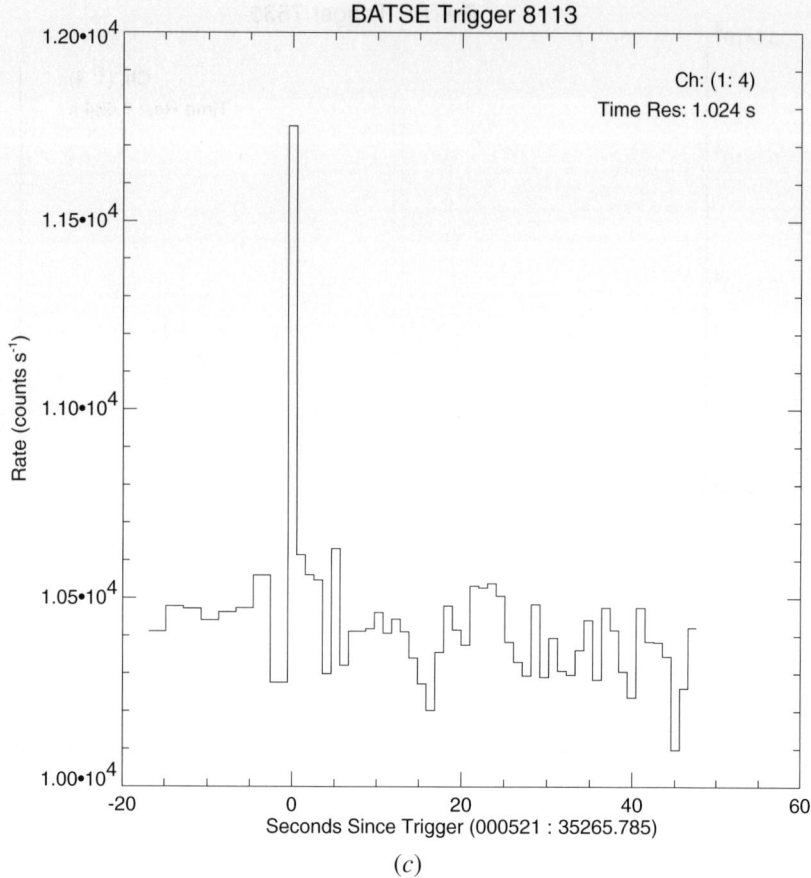

Figure 13.1. (Continued.)

included detectors with sensitivity to such delayed emission. As it turned out, the detectors were almost exactly matched to the cosmic GRB phenomenon.

The gamma-ray detectors were, of necessity, quite small. Each spacecraft had six 10 cm^3 CsI scintillation detectors distributed so as to give equal sensitivity in all directions. The energy range covered was 0.2–1.0 MeV (later to 1.5 MeV). The data-recording system was triggered if there was a rapid rise in counting rate above the background. The counting rates were recorded in increasing logarithmetric time intervals so that the time characteristics of the complete burst could be recorded.

Although the instruments, which were built at the Los Alamos Scientific Laboratory and the Sandia Laboratories in Albuquerque, New Mexico, were not classified, there was a general air of secrecy about the missions. The astronomical gamma-ray community (small at that time) were not generally aware of their

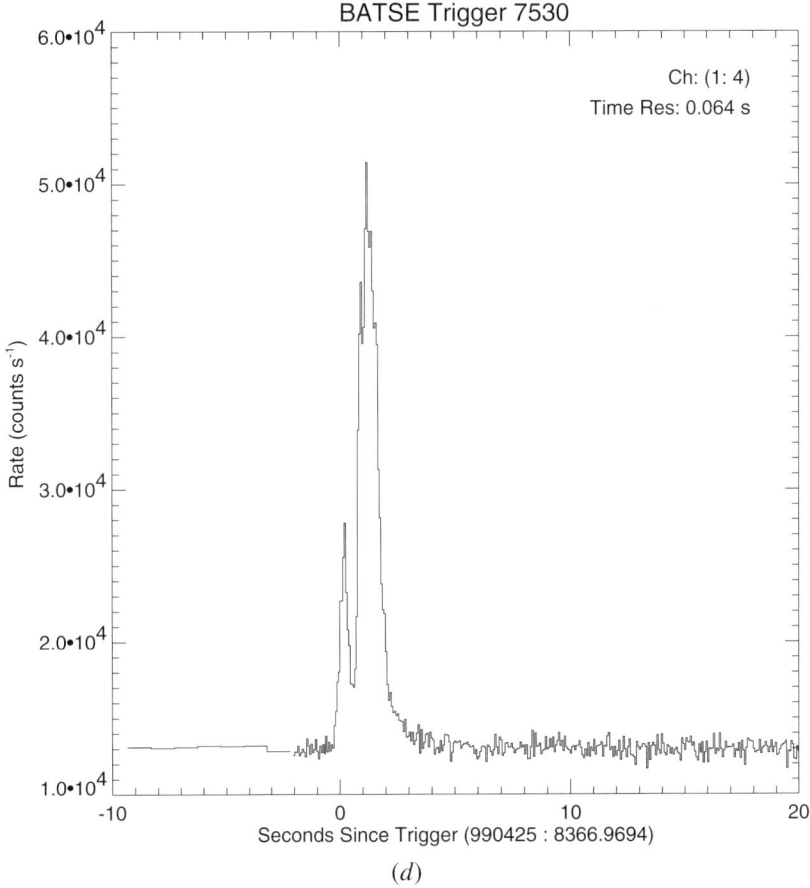

Figure 13.1. (Continued.)

existence. In practice, of course, there was no *a priori* reason to expect the existence of cosmic GRBs so the Vela satellites would not have been judged to be of much astronomical interest.

In the discovery paper [14], the detection of 16 bursts by the Vela 5 and Vela 6 series of satellites was announced. In a rather prosaic style, which did not belie the significance of the findings, the essential parameters of the phenomena were described. The bursts, which were detected between July 1969 and July 1972, had durations from 0.1 to 30 s and fluences from 10^{-5} to 10^{-4} erg cm^{-2} (these were very bright by today's standards when fainter bursts are detected at a daily rate). The fluence is the total energy detected from the burst, integrated over its duration. The time structure within each burst was different and no discernible pattern was apparent. Their arrival directions, which were only

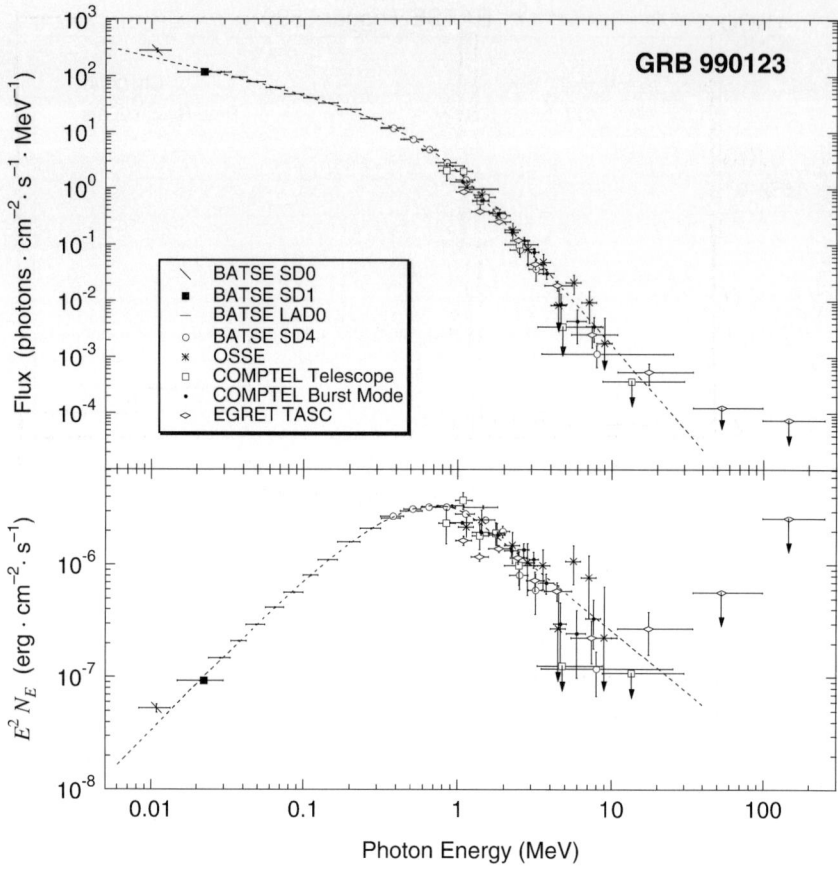

Figure 13.2. Complete spectrum of GRB990123 as measured by BATSE, OSSE, COMPTEL, and EGRET on CGRO [7]. (Figure: M Briggs.)

roughly determined (to within 10°), could not be identified with any one direction, with the direction of the sun or earth, or with that of recent supernovae. The long delay between the detections and the publication was not that the subject matter was classified but that the experimenters had to satisfy themselves that the triggers were real. They were not, after all, astronomers and the detection of an astronomical phenomenon did not rank high on their list of priorities. It took some time for the experimenters to convince themselves that the bursts were truly of cosmic origin. However, immediately after publication, Goddard gamma-ray astronomers [8], using the hard x-ray detector on board IMP-6, confirmed the phenomenon and showed that the spectrum peaked between 100 keV and 1 MeV. Their experiment was designed to study flares on the sun and the investigators were unlucky not to have discovered the GRBs first.

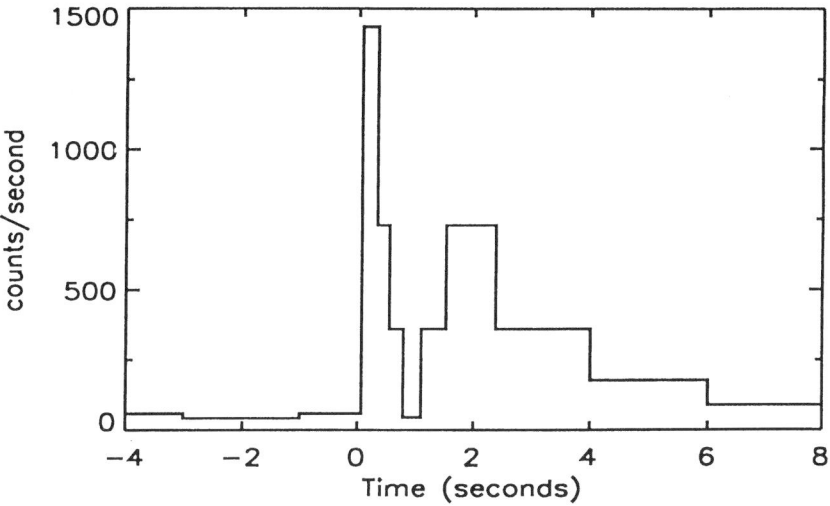

Figure 13.3. First GRB as recorded by Vela 4a [5].

Subsequent analysis by the Vela team showed that, in fact, a burst had been detected by detectors on the Vela 3 and 4 series of satellites on 2 July 1967; this was really the first gamma-ray burst ever detected (figure 13.3).

13.3 Properties of gamma-ray bursts

13.3.1 Time profiles

Bursts are detected with durations from milliseconds to thousands of seconds in duration. The pulse shapes are highly irregular, some with very smooth light curves, others with detailed time structure. Various attempts have been made to sub-divide the bursts by their durations; there is evidence for at least two classes of bursts with a break between the two populations at $t = 2$ s (figure 13.4). There is sub-structure within the bursts but no consistent pattern; there is no evidence for periodic structure. Rise-times of the overall burst and of substructures within the burst tend to be sharper than fall times. At high energies (>30 MeV), there is evidence for a delayed emission but this may be from a different component in the emission process (see later).

13.3.2 Energy spectra

There is not nearly as wide a variation in behavior with photon energy as there is with time. Almost all the power is emitted at energies above 50 keV. The spectra show no spectral lines and is generally a smooth continuum: the distribution of peak energies is surprisingly narrow. The spectra can be represented by two power

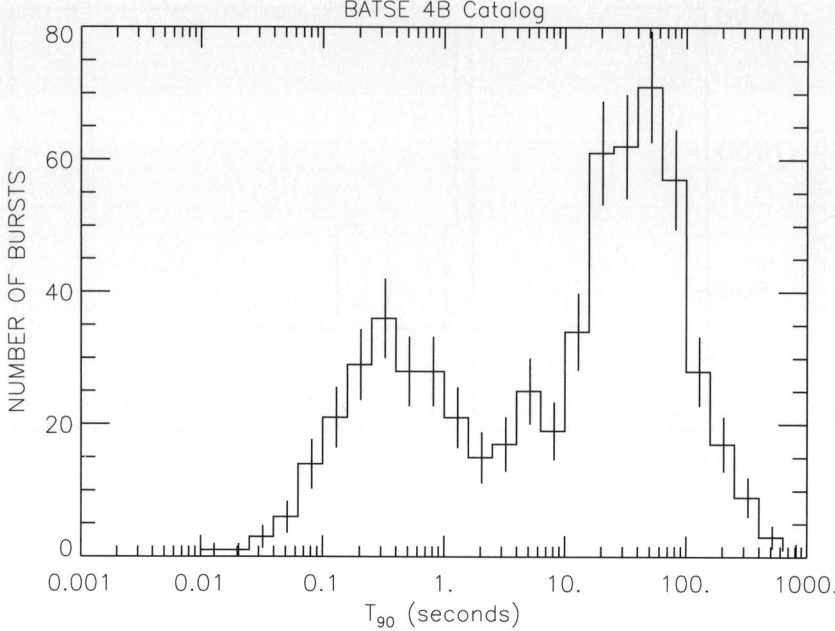

Figure 13.4. Time duration distribution of GRBs as recorded by BATSE. The duration used here is 'T_{90}' which is the interval of time during which the GRB emits 5% of its energy to that when it emits 95%. The counts recorded in the BATSE detectors are assumed proportional to the total energy emitted. (Figure: http://www.batse.msfc.nasa.gov/batse/grb/)

laws, with differential index from 0 to -1.5 up to the power maximum, and -2 to -2.5 thereafter (figure 13.2). There is some evidence for a change of spectrum during the burst with most bursts softening over their duration.

13.3.3 Intensity distribution

If the distribution of GRB sources is homogeneous in space, i.e. if the density and luminosity are independent of position within the observable volume, then the integral intensity distribution will be $N(> F) \propto F^{-3/2}$. The observed distribution is shown in figure 13.5 as a full line; the $-3/2$ power law is shown as a dotted line. There is a clear deviation at lower intensities, possibly indicating that the edge of the distribution in space has been reached.

13.3.4 Distribution of arrival directions

With more than 2700 GRBs detected by BATSE with positions good to a few degrees (figure 13.6), the most surprising aspect of the phenomenon becomes

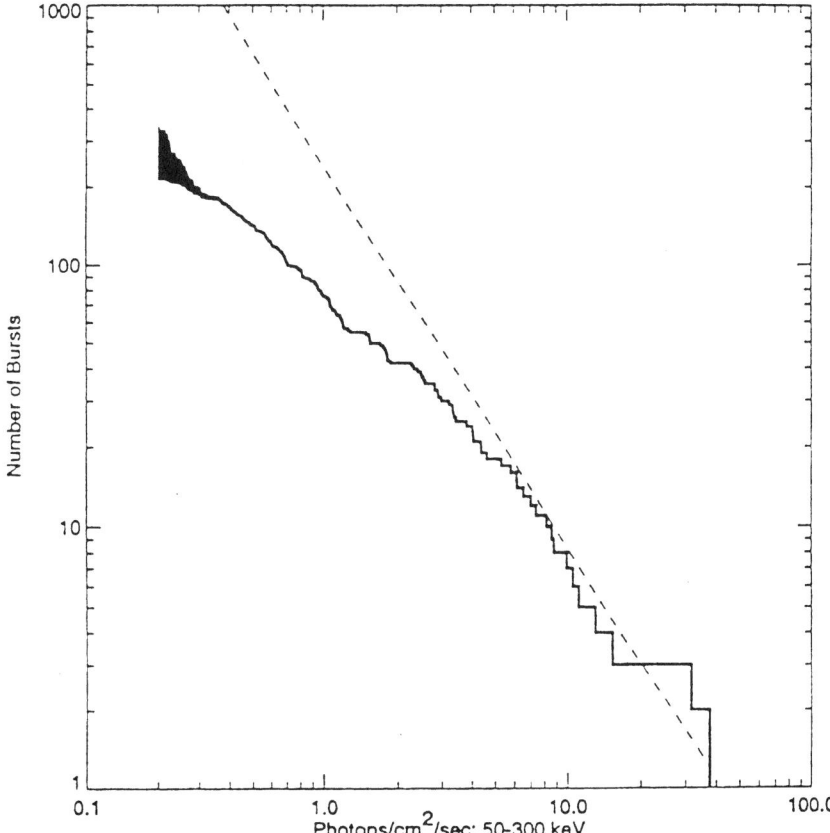

Figure 13.5. Intensity distribution of GRBs showing deviation from the −3/2 power law expected if GRBs are homogeneously distributed in Euclidean space [11]. (Reprinted from Fishman 1995 *Publ. Astron. Soc. Pac.* **107** 1145. Copyright 1995 Astronomical Society of the Pacific; reproduced with permission of the editors.) (Reproduced with permission of *PASP*.)

apparent; they are completely isotropic in two-dimensional coordinates. Even the weakest bursts show no evidence for clustering; there is no evidence for repeated bursts, no correlation with any known class of cosmic object, no evidence for a dipole or quadropole moment. For the bursts detected by BATSE, the limit on the dipole moment is -0.014 ± 0.014 and quadropole moment 0.004 ± 0.007.

13.4 The location controversy

From the beginning, GRBs were perfect objects for theoretical speculation. With no distance measurements, they could be as close as the Oort Belt (a distribution

2704 BATSE Gamma-Ray Bursts

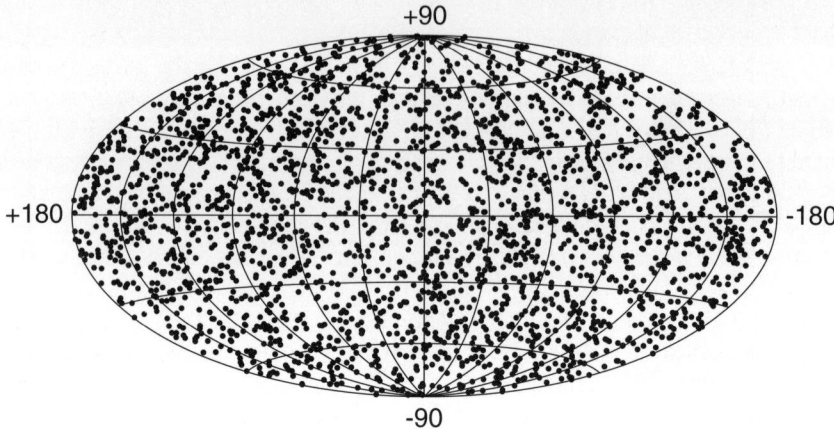

Figure 13.6. Distribution of the arrival directions of the 2704 GRBs detected by BATSE in galactic coordinates. The grey levels are a measure of the fluence. (Figure: http://www.batse.msfc.nasa.gov/batse/grb/)

of small solid objects outside the Solar System) or as distant as the farthest quasars. There was no shortage of suggestions as to their origin—for a time the number of theories exceeded the number of bursts detected. As always in astrophysics, the fewer the observable facts, the wilder the speculation becomes. Primordial black holes, anti-matter comets, starquakes on neutron stars, flares on AGN: all had their champions. By the mid-1980s the controversy as to the origin of the bursts had narrowed into two camps: those who believed that they were associated with neutron stars in the Galaxy (the majority); and those who believed they were compact objects at cosmological distances (the minority). The energy budget increased dramatically with distance and, hence, the difficulty of explaining the energy production.

That the distribution of GRB positions in the sky was apparently isotropic and that the intensity distribution did not follow a 3/2 power law became very apparent in the early years of the CGRO mission with the huge detectors of BATSE detecting bursts at the rate of one per day. The conundrum was simple: if we were seeing to the edge of the distribution (as the intensity distribution showed) and if the distribution was uniform in all directions, then we must be at the center of the distribution. This was not the first time that mankind was presented with an egocentric view of the universe.

The typical distance to the burst sources was critical to understanding the scale of the phenomenon and, thus, to postulating the nature of the burst emission mechanism. If they were close, e.g. within the Solar System, then the energetics became trivial. If within the Galaxy, the problem was not so simple but not beyond the bounds of the physics of compact objects. If the sources were extragalactic and at cosmological distances, then the energy problem became extreme and could hardly be contemplated.

The lack of detectable curvature in the arrival times of the plane wave of gamma rays in the bursts (when more than three interplanetary satellites were involved in the detection) eliminated an origin within the Solar System (also it was hard to believe that such an energetic phenomenon would not have been noticed at other wavelengths). The sun is significantly displaced from the galactic center; hence, if the GRB sources were isotopically distributed within the Galaxy we would expect to see a distribution that peaked in the direction of the galactic plane and towards the galactic center. The absence of such an anisotropic distribution could only be accommodated if the distribution was centered on the Galaxy but spread out in a wide halo, so distant that their differential distance to the sun and to the galactic center became unimportant. No known component of the Galaxy has such a wide distribution.

Prior to the launch of the CGRO, there was an increasing consensus that GRBs were associated with neutron stars (the exact mechanism for gamma-ray emission was in dispute but there was no shortage of possibilities) and that the conclusive proof would come when BATSE revealed the anisotropic distribution of locations which would clearly point to a galactic origin. Either the underlying structure of the galactic plane would be seen in the GRB distribution or, if there was no such structure and a galactic halo was implied, the contribution from the GRBs in the equally large halo in the nearby galaxy, M31, would be apparent. The neutron star hypothesis was strongly supported by the apparent detection of cyclotron lines in the spectrum of bursts: these emission and absorption features might be expected in the strong, but uniform, magnetic fields of neutron stars.

But in the first year of operation of BATSE, the distributions were found to be isotropic and there was no evidence of cyclotron lines. The debate about the location of the sources of GRBs heated up, with the community now evenly divided between those who favored the extended galactic halo origin and those who insisted on a cosmological distribution. A flurry of publications, workshops, and symposia failed to resolve the issue. A formal debate (to commemorate the 75th anniversary of the famous Curtis–Shapley debate on the location of the spiral nebulae) was held in Washington, DC in 1995 (see historical note: the great debate). The matter was not to be resolved there and had to await new data, in this case the discovery of the long-awaited lower-energy counterparts to the bursts.

13.5 Counterparts

Few astronomical phenomenon are confined to emission in a single waveband. Emission at other wavelengths is particularly important for gamma-ray studies because of the inherently poor angular resolution of gamma-ray telescopes. If newly discovered sources are to be identified with known astronomical objects, the source location must be refined to minutes of arc. This is rarely possible with gamma-ray observations alone. The same situation pertains to GRB locations. However, because of the inherently short time structure in the bursts, it is possible to locate the bursts using differential timing between widely separated spacecraft. By placing small burst detectors on deep space planetary missions, an interplanetary network was set up in the late 1970s that is capable of locating sources to a few arc-minutes. Unfortunately, this could not be done in real time so that it was often a few days before source positions were determined and the counterparts sought. In this case the assumption was that in addition to possible emission simultaneous with the gamma-ray emission ('prompt emission'), there was an additional component which lasted an indeterminate time after the burst ('delayed emission'). Although several possible associations were found, these were not convincing and at the time of the launch of BATSE there was no GRB with a counterpart reliably identified.

After the launch of the CGRO in 1991, BATSE demonstrated its ability to detect weak GRBs at a rate of one per day. It had the capability to locate bursts within a circle of a few degrees radius within a few hours after the burst detection. This was not sufficient to give a convincing detection of counterpart radiation during the burst. In 1994, a new system went into play. Using BATSE data intercepted in the first 2 s of the burst at the Goddard Space Flight Center, the BACODINE (BAtse COordinate DIstribution NEtwork) determined the source coordinates and distributed them world-wide over the internet within five seconds of the burst occurrence [3]. Suddenly it was possible to make almost real-time observations of GRB counterparts from ground-based optical, infrared, and radio telescopes. However, the positional accuracy was still quite poor. In addition, some large ground-based telescopes took a finite time to disengage from their regular observing program and to trundle across to a new observing target. Several small wide-field telescopes were built to respond specifically to the GRB alerts. Despite this intense effort, no counterparts were found using BACODINE until 1998 [1] when GRB990123 was detected by the robotic ROTSE telescope. The charge coupled device (CCD) images showed that at the time of the GRB, the brightness was $m \approx 11.7$ and that it increased by three magnitudes over the next minute and then faded slowly. No other GRB has exhibited optical emission at this level and GRB990123 may be an unusual case.

The identification breakthrough came when the Italian–Dutch satellite, BeppoSAX, with an unusual complement of instruments [4], was launched in 1996. Not planned primarily to detect GRBs and largely ignored by the astronomical community, this Dutch–Italian mission proved to have the decisive mix of telescopes to fix the source positions and to permit the resolution of the source distance scale. These consisted of a small, all-sky, GRB monitor, GRBM (40–700 keV), two wide-field x-ray cameras (2–26 keV) which together covered 5% of the sky, and a series of narrow-field x-ray cameras (0.1–300 keV); by redirecting the orientation of the satellite the narrow-field telescopes could be directed to an interesting area of the sky.

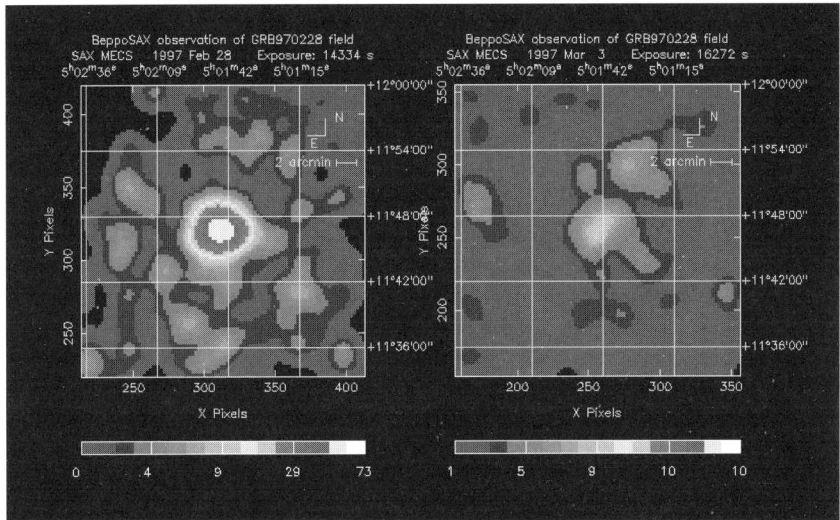

Figure 13.7. The first x-ray counterpart recorded by BeppoSAX. On the left: x-ray image of GRB970228 position eight hours after burst. On the right: same region, two days later. Reprinted from Costa *et al* 1997 *Nature* **387** 783 with permission of Nature Publishing Group.)

The first hint that the long saga of the search for the GRB distance scale was over came with the detection of GRB970228 by BeppoSAX[9]. This GRB fell within the field of view of one of the wide-field cameras so the prompt x-ray source could be located with a positional accuracy of 3 arc-min radius; within 8 hr the narrow-field cameras were slewed to that position and the source intensity was monitored as it declined (figure 13.7). The existence of afterglow x-ray radiation had been predicted in some theoretical models of GRBs at cosmological distances but had not been detected previously. The location was further refined so that deep-field searches could be made with optical and radio telescopes. An optical source with magnitude $R = 21$ was detected which faded to $R = 23.3$ within six days. An accurate redshift could not be determined but, as the transient source faded, it appeared to be superimposed on an extended faint source. A plausible hypothesis was that this was a distant galaxy and that the burst had originated in that galaxy.

Conclusive evidence that the GRB sources were at cosmological distances came quickly. GRB970508 was detected by BeppoSAX with the same sequence of observations and the optical counterpart this time was found to be of magnitude $R = 19.7$. Although it faded by two magnitudes over the next ten days, it was sufficiently bright to yield an absorption redshift of $z = 0.835$. A flaring radio source was also detected.

Within 18 months, nine host galaxies had been identified with magnitudes ranging from $23 < R < 27$. The size, morphology, and color of these suggested that they were star-forming galaxies. Not all had measurable redshifts but those that were measured were in the range $z = 0.83$–3.4.

The question as to whether GRBs and conventional supernova explosions are related is a controversial one. There was a positional coincidence of GRB980425 and SN1998bw which was at a redshift of 0.008. If the two were related, then this would be the closest GRB whose distance had been measured. It would be expected, therefore, that it would be very bright; it was not, which indicated that if the association was real, this was not a standard GRB (although it had already come to be accepted that GRBs were not the standard candles that they were once believed to be). In fact, SN1998bw is an unusual supernova but it is still a mystery as to why the GRB intensity should be a factor of 10^4 less luminous than expected based on that distance.

13.6 The high energy component

It was not anticipated that EGRET would be particularly sensitive to GRBs as the GRB properties were known before the mission launch. In fact, prior to the launch of CGRO, the highest energy photons observed from a burst had an energy of 80 MeV. Because the EGRET spark chamber had an inherent deadtime which lasted from about 600 ns to 100 ms after a trigger, it was impossible to measure the total flux of MeV gamma rays in a burst. The combined collection area of the BATSE detectors was 1.5 m^2; that of EGRET, for most energies, was <0.15 m^2. Assuming a power-law spectrum that had a negative exponent >2.0, the number of photons within the EGRET energy range would necessarily be small. Hence, not much emphasis was given to the potential for EGRET to contribute to the investigation of GRBs. Only the very brightest would be expected to give a signal and it would have been no great surprise if the spectrum was found to turn over below 30 MeV.

In addition to the spark chamber, EGRET had a sodium iodide calorimeter (chapter 3) which could detect GRBs independently of the spark chamber either as an independent trigger or in response to a command from BATSE [10]. Compared with the spark chamber, the calorimeter or Total Absorption Shower Counter (TASC) had a lower energy response, a wider acceptance angle, and no deadtime. This proved to be very effective for measuring the HE energy spectra of 15 GRBs. These measurements did not show any evidence for a break or cut-off in their spectra but showed a range of hard spectral indices. The TASC provided virtually no directional information. In addition, the anti-coincidence shield around EGRET was sensitive to x-rays in the 25–50 keV range.

Both the EGRET spark chamber and the TASC detected the bright BATSE GRB910503. Although only nine photons were seen in EGRET, the detection was significant and showed that EGRET had a contribution to make to the study of GRBs. In fact, all of the very bright bursts which occurred during the mission

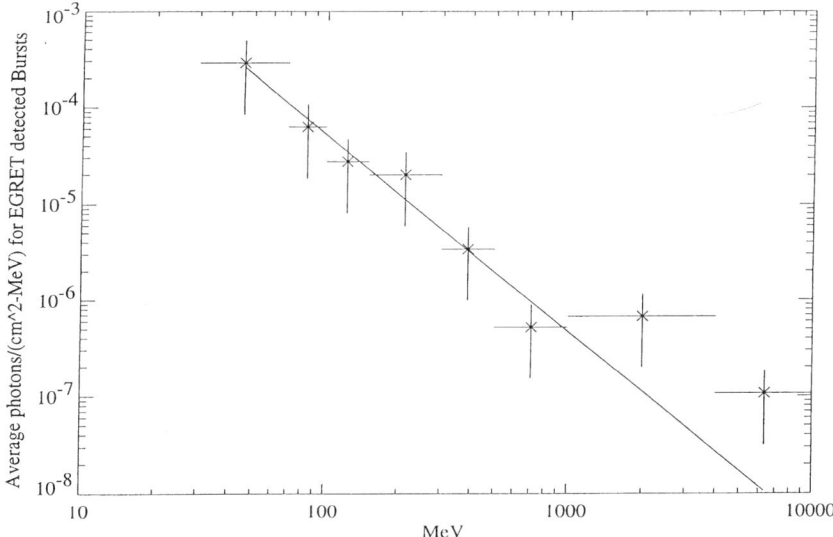

Figure 13.8. Average spectrum of four GRBs detected by EGRET over the 200 s from the start of the BATSE signal. The differential photon spectral index is -1.95 ± 0.25 [10].

within $\pm 30°$ of the axis of EGRET were detected. This only amounted to five bursts but, taken together, some definite information can be gleaned. The combined spectrum is shown in figure 13.8. The data, taken over the first 200 s after the BATSE burst onset, can be fitted with a power law with differential spectral index -1.95 ± 0.25. There is no evidence for a cut-off up to the highest energy for which there is meaningful statistics. Since there is deadtime in the EGRET readout, the absolute flux is not meaningful but it is clearly higher than that recorded.

A unique feature of the EGRET GRB detections was the first evidence for delayed gamma-ray emission, the so-called afterglow. Three of the five EGRET bursts showed evidence for emission well after the BATSE burst had disappeared. The most dramatic event was the burst, GRB940217, which was detected by BATSE as a bright burst (fluence of 7×10^{-4} erg cm^{-2} of duration 180 s). During this 180 s, EGRET detected a total of 10 photons from the direction of the burst. In the 1.5 hr after the burst, a further 18 HE photons were detected (4.7 would have been expected in this time interval from the background). The distribution of the EGRET-detected photons with time and energy is shown in figure 13.9 with the BATSE burst superimposed. The post-burst observation was severely curtailed as the EGRET field of view was occulted by the earth. However, there was enough information to show that 50% of the delayed emission was in photons with energy greater than 30 MeV (compared with only 2% in the prompt burst). Delayed emission was thus predominantly a high energy feature of the burst. In

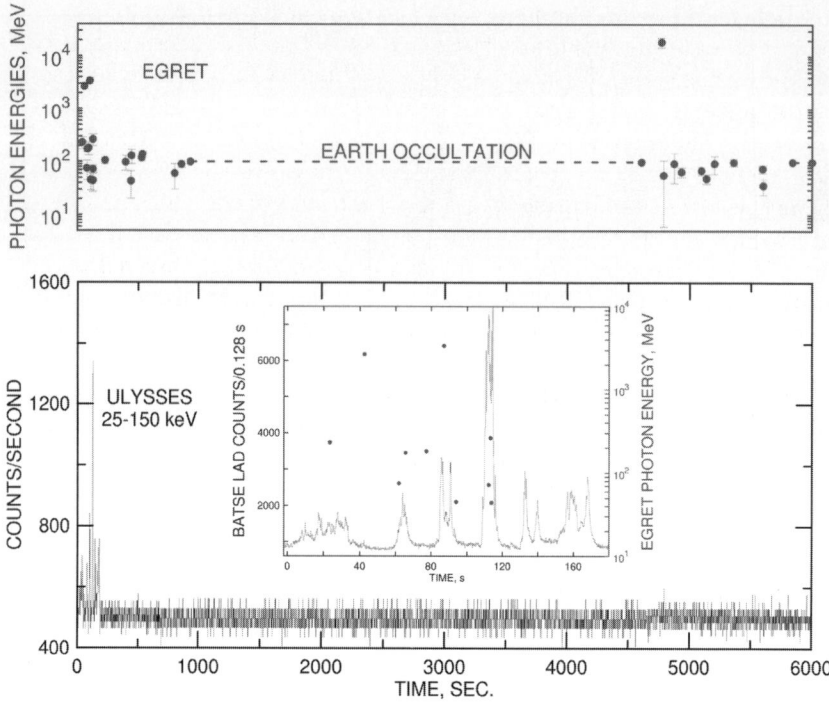

Figure 13.9. (Top) energy of individual photons detected by EGRET plotted as a function of time since the onset of the BATSE GRO 940217 burst. (Bottom) the counting rate recorded on the Ulysses x-ray detector [13]. (Reprinted from Hurley *et al* 1994 *Nature* **372** 652 with permission of Nature Publishing Group.)

summary, the EGRET observations were consistent with *all* GRBs having a hard component.

Most remarkable of the EGRET GRB results was the detection of one photon of energy 18 GeV from the GRB940217 direction, 1.5 hr after the BATSE detection. Seldom has the detection of a single photon caused such excitement. Normally the detection of only one photon would not be accorded much notice but, in this case, a photon of this energy was detected only once in every two weeks of EGRET operation; its detection coming from the burst direction is statistically significant. With the detection of a photon of 18 GeV it is no longer a forlorn hope that the GRB spectrum will extend to GeV, and possibly even to TeV energies, and, therefore, into the realm of VHE ground-based detectors.

The detection of HE photons immediately begs the question as to where the spectrum breaks. Since the distances are cosmological, there must be a steepening somewhere above 100 GeV due to intergalactic pair production on infrared photons (chapter 14). But the atmospheric Cherenkov techniques are sufficiently sensitive that the detection of just 3–4 photons from the GRB position would be statistically significant so that a very small fluence is detectable. There have been a number of attempts to search for VHE (>300 GeV) components using ground-based ACT and particle air shower detectors. Since the ACT instruments are narrow-field instruments, there is little chance of having the GRB within the field of view of the telescope during the prompt phase. Particle arrays have fields of view of more than a steradian and operate continuously; although they are less sensitive they may have the GRB position in their sight as the GRB occurs. The Milagrito experiment (prototype of Milagro, chapter 2) observed 54 such positions and from one of them, GRB970417a, observed a signal during the GRB duration of 7.9 s that was significant at the 10^{-3} level when all trials are allowed for [2]. This is the strongest evidence for the detection of gamma rays of energy >500 GeV from a GRB. This was a weak burst in BATSE and no positional information was available. There is no obvious reason why this GRB, and no other, should be detectable at VHE energies.

If real, the emission of photons of such high energy and so long after the prompt burst is a real challenge to burst models. It is not clear how, if at all, this delayed VHE gamma-ray emission is related to the delayed emission in the x-ray and optical counterparts (although in some models it appears as a Compton component).

13.7 The afterglow

The first hint of the existence of an afterglow, the secondary component, that follows in the aftermath of the GRB, came from HE gamma-ray observations as discussed earlier. The observations of x-ray emission by the BeppoSAX satellite firmly established the existence of an x-ray afterglow and led to the discovery of optical and radio afterglows. These observations were made possible by the rapid dissemination of the BeppoSAX GRB positions via the Gamma-ray burst Coordinates Network (GCN) which succeeded the BACODINE system. Unlike the BATSE positions, which were available within seconds of the burst detection, these coordinates were not available until six hours after the GRB occurred.

When a good position for the GRB is established, it is possible to identify the galaxy in which the GRB source is located. This is important as it gives some information as to the nature of the source. The GRB sources appear to be in star-forming regions, not far from the center of the galaxy; the galaxies are blue and not very bright.

The x-ray afterglow is observed to decay with time, t, as $F_x \propto t^{-(1-1.5)}$. The spectrum is softer than that in the prompt emission. Only in half the cases where a definite x-ray afterglow was measured is there detectable optical emission. This is always faint ($\approx 20 m_v$) and decays rapidly ($F_0 \propto t^{-(0.8-2.2)}$). Only for GRB990123 was there really bright optical emission detected [1]. However, this should more properly be considered prompt emission. The absence of optical emission is clearly significant and several theories have been advanced to explain these 'dark' GRBs. The radio afterglows are particularly interesting because the rapid variability (scintillation on passing through the interstellar medium) is observed initially and this sets a limit to the maximum angular size of the afterglow region. Typically, the scintillation ceases about a month after the burst. For the sources with measured z, it is then possible to deduce the linear dimension of the source and, knowing the time since the GRB occurred, the velocity of the expanding fireball. It is these data that are used to calculate the total energy associated with each GRB where beaming is assumed [12].

13.8 Models

Proposed explanations of the GRB phenomenon are some of the hottest topics in astrophysics and no field is changing so rapidly. Hence, any attempted summary of current thinking will probably date very rapidly. There is an emerging consensus on some aspects of the problem. Whereas the initial models strove only to explain the observed prompt gamma-ray properties, current theoretical activity has centered on explaining the afterglow.

13.8.1 Central engine

As with AGN, it is possible to consider the radiation that constitutes the GRB without fully defining the central engine. The current paradigm is that of a relativistic fireball [15] that originates from a catastrophic event involving the formation of black hole. Three scenarios have been proposed:

(1) the merging of two neutron stars to form a black hole;
(2) the core collapse of a massive star (10 $M_\odot$, a so-called hypernova);
(3) a supranova, the collapse of a neutron star into a black hole formed in a supernova explosion.

The energy could be extracted from the black hole as it is postulated to happen in an AGN, i.e. the rotational energy is tapped via the unipolar inductor mechanism.

13.8.2 Total energies

With distances now measured to over 20 GRBs from afterglow identifications, it is possible to estimate the total energy emerging from the source if isotropic emission is assumed. These estimates show that the total energy ranges over three decades: from 5×10^{51} to 3×10^{54} erg. This is more energy than is emitted in any other astrophysical phenomenon since the Big Bang. It is not surprising that some theorists clung to the hope that GRBs would have a galactic explanation. Also as discussed earlier, the time scale of the emission is very small so that the dimensions of the emitting regions must also be small, e.g. a 3 ms flare indicates an emission region of 10^8 cm. The emission of so much energy in a small space is an immediate problem; with such a high density, the photon–photon pair-production process would have such a large cross section that it would be impossible for the gamma rays to escape.

13.8.3 Beaming

Because of these considerations there is a growing consensus that GRBs must be beamed. The same arguments that were used to explain the high energy emission from blazars (chapter 12) are now applied to GRBs. This argument applies no matter what the central engine is. If there is beaming, then the total energy is considerably reduced, the absorption problem is reduced, and the ubiquitous bulk motion in a relativistic jet is invoked. Whereas the bulk Lorentz factor, Γ, in AGN had values of approximately 10, to explain GRBs, Γ must have values in excess of 100. In this way the total energy emitted from the GRB can be reduced by a factor of 10^{2-3}, i.e. to 10^{51} erg, comparable to that of a supernova explosion. To some extent this displaces the primacy of GRBs as energy sources but they still have pride of place for the *rate* of energy emission. A corollary to this explanation is that GRBs must be much more frequent than was previously supposed. If the beam factor is 1/500, then the real rate of GRBs may be $\approx$500/day, instead of the observed 1/day because the emission is only visible within the solid angle of $1/\Gamma$. However, the effect of collimation (see later) must also be taken into account.

13.8.4 Emission mechanism

The most popular explanation for the emission of the gamma radiation in GRBs is that they are the result of a relativistic fireball [15]. Without considering what causes the fireball, it is postulated that shells of relativistic material are emitted into the interstellar medium in a succession of explosive events. As different shells interact, they produce shocks. The GRB is thought to be caused by these internal shocks. The afterglow is due to the external shock, the termination of the jet as it interacts with the surrounding interstellar medium. In this scenario the source of the emission for both the prompt and delayed emission is the relativistic outflow in the jet and it occurs in regions that are optically thin so the gamma rays can escape. The observed x-ray, optical, and radio afterglow is synchrotron emission from the relativistic electrons.

13.8.5 Geometry

The beaming models have received a major boost from their ability to explain the measured break in the afterglow luminosity. This break can be used to ascertain the geometry of the source and to determine the total energy in the jet. If a blob of relativistic matter is radiating isotropically in its rest frame and is moving relativistically, the radiation in the observer's frame is beamed into a cone of angle $1/\Gamma$. Initially to the observer there is no difference between emission from a uniform expanding sphere or from a jet with finite opening angle, θ. The bulk motion will gradually decrease and the opening angle will exceed that of the jet, $1/\Gamma > \theta$. As Γ decreases, the observer sees more of the beamed $(1/\Gamma)$ emission so the decay in observed flux seems slower. Once this angle exceeds θ, the observer sees the emission in its entirety and the true decay constant is seen. It is at this point that there is a break in the light curve. Using this relationship it was possible to deduce the opening angle for 15 GRBs with observed afterglow light curves [12]. The total energy could then be estimated. It was found that for these GRBs the total energies were all approximately 5×10^{50} erg. The small spread in total energy was surprising as it indicated that the central engine in all GRBs was remarkably similar. This small spread favored models in which the central engine was a unique catastrophic event (such as a hypernova). A corollary of this geometrical explanation is that there should be frequent observations of 'orphan' GRBs at longer wavelengths in which the afterglow (at wide angles) is observed but the narrow angle GRB emission is missed.

An interesting side product of the fireball model is that the mechanism could, in principle, produce ultrahigh energy cosmic rays. Thus, in one fell blow, two outstanding astrophysical mysteries could be solved by a single phenomenon [16]. However, the evidence for the association is not conclusive.

Historical note: the great debate

Controversies in science are seldom solved by public debate. The 'Great Debate' that was held in 1995 in Washington DC on the distance scale to GRBs was no exception [6]. Still it was a historic occasion and that the event was held at all highlighted the importance attached to the issue in the astronomical community. Designed to celebrate the Diamond Jubilee of the famous debate between Curtis and Shapley on the nature/distance scale of the spiral nebula (which was equally inconclusive), the event attracted much attention. Like the 1920 debate it was held in the Main Auditorium of the Smithsonian Museum of Natural History and, hence, the event was replete with history. After a historical reprise of the 1920 debate, Jerry Fishman, the Principal Investigator for the BATSE team, gave a summary of the observational properties of the GRBs. Then the principal exponents of the two schools of thought concerning the distance scale argued their cases: Don Lamb, of the University of Chicago, for a distant galactic halo scale; and Bohdan Paczynski, of Princeton University, for the cosmological scale. The audience seemed equally split and, despite the excellent presentations, few opinions were changed. As one commentator, writing a year later, put it: 'The event was enjoyable, educational, inspiring and even fun. Unfortunately the distance scale to gamma-ray bursts is still unknown!'. It is a measure of the importance of observation to astrophysical thought that it only took one small set of observations of GRB counterparts to decide the issue conclusively and this was only a few years later.

References

[1] Akerlof C *et al* 1999 *Nature* **398** 400
[2] Atkins R *et al* 2000 *Astrophys. J. Lett.* **533** L119
[3] Bartelmy S D *et al* 1998 *Proc. 4th Huntsville GRB Symp. (AIP Conf. Proc. 428)* ed C A Meegan, R D Preece and T M Koshut (New York: AIP) p 99
[4] Boella G *et al* 1997 *Astron. Astrophys. Suppl.* **122** 299
[5] Bonnell J T and Klebesadel R W 1996 *Conf. on Gamma Ray Bursts (Huntsville, AL, 1995)(AIP Conf. Proc. 384)* ed C Kouveliotou, M F Briggs and G J Fishman (New York: AIP) p 977
[6] Bonnell J T, Nemiroff R J and Graziani C J 1996 *Conf. on Gamma Ray Bursts (Huntsville, AL, 1995) (AIP Conf. Proc. 384)* ed C Kouveliotou, M F Briggs and G J Fishman (New York: AIP) p 973
[7] Briggs M S *et al* 1999 *Astrophys. J.* **524** 82
[8] Cline T L *et al* 1973 *Astrophys. J. Lett.* **185** L1
[9] Costa E *et al* 1997 *Nature* **387** 783
[10] Dingus B, Catelli J R and Schneid E J 1998 *Gamma Ray Burst 4th Huntsville Symposium (AIP Conf. Proc. 428)* ed C A Meegan, R D Preece and T M Koshut (New York: AIP) p 349
[11] Fishman G T 1995 *Pub. Astron. Soc. Pac.* **107** 1145
[12] Frail D *et al* 2001 *Astrophys. J. Lett.* **562** L55
[13] Hurley K *et al* 1994 *Nature* **372** 652
[14] Klebesadel R W *et al* 1973 *Astrophys. J. Lett.* **182** L85
[15] Meszaros P and Rees M 1997 *Astrophys. J.* **476** 232
[16] Waxman E 1995 *Phys. Rev. Lett.* **75** 386

Chapter 14

Diffuse background radiation

14.1 Measurement difficulties

From the cosmologist's perspective, there is no aspect of gamma-ray astronomy as important as the truly diffuse gamma-ray background. The extremely good penetrating power of HE gamma rays raises the prospect of gamma rays coming from the earliest times (out to $z = 100$) and, hence, giving a glimpse of conditions at cosmological times that is otherwise not available. VHE gamma rays are of little use in this regard since the techniques available do not permit the detection of a diffuse background and are unlikely to do so in the future. Also VHE gamma rays do not have the penetrating power of HE gamma rays—ironically this can be used to provide unique information on the cosmic infrared background out to distances of $z = 1$ (see later).

The diffuse background exists at all wavelengths but, in different wavebands, it has different origins. Much of it originates in thermal processes (microwave to ultraviolet) with the radio, x-ray, and gamma-ray backgrounds arising from non-thermal processes.

It is notoriously difficult to make measurements of the diffuse background at any wavelength. This is particularly true of gamma rays. To establish that a flux is truly diffuse and extragalactic, it is necessary to eliminate the possibility that the measured flux is:

- instrumental,
- a foreground diffuse source,
- the sum total of unresolved weak discrete sources, and
- the residue of the diffuse galactic flux.

Instrumental backgrounds are difficult to eliminate, particularly at gamma-ray energies, where secondary gamma radiation can be produced by the interaction of the cosmic ray background with the space craft components. It is difficult to simulate such backgrounds although careful design can minimize these effects. The presence of a 'diffuse' foreground can be inferred from other measurements, e.g. the zodiacal light contribution to the infrared background, and subtracted. The unresolved point-source contribution is important in itself and is related to the total population of known types of source extrapolated to greater distances. The subtraction of the galactic contribution depends critically on how well the model can be fitted to the observed well-defined galactic flux.

14.2 Diffuse gamma-ray background

14.2.1 Observations

The CGRO mission has provided the best measurements of the diffuse gamma-ray spectrum from energies of 1 MeV to 100 GeV [11]. These data came from the COMPTEL and EGRET telescopes, both of which had exceedingly low instrumental backgrounds and which have provided consistent results in the energy range where they overlapped (30 MeV).

The SAS-2 mission provided the first direct evidence for the existence of a diffuse component above 30 MeV. It had a lower instrumental background than the longer COS-B mission. The SAS-2 measurements indicated that the background could be represented by a power law with spectral index -2.35 ($+0.4, -0.3$). EGRET has provided more definitive measurements from 30 MeV to 10 GeV [12] where the data can be fitted with a power law spectral index of -2.10 ± 0.03 (figure 14.1). Although the data are sparse at higher energies, it is in agreement with a smooth extrapolation up to 120 GeV. The integral flux above 100 MeV is $(1.45 \pm 0.05) \times 10^{-5}$ photons cm^{-2} s^{-1} sr^{-1}. When the galactic plane is excluded, it appears that this radiation is completely uniform across the sky, i.e. when the sky is divided into 36 equal bins (excluding the plane), the variation between bins is $<20\%$ and compatible with the statistical accuracy of the measurements.

The measurement of the diffuse extragalactic background is an important one and depends critically on the EGRET observations and their interpretation, in particular the success in eliminating any possible contribution from other sources. The confirmation of the measurement by GLAST will be one of the highlights of that mission.

14.2.2 Interpretation

If it is accepted that the observed diffuse flux is from beyond the Galaxy, i.e. it is definitely of extragalactic origin, then the mechanism of emission must be considered. The high degree of isotropy eliminates any possibility that it is associated with the Local Group or the Virgo Cluster, and that, therefore, it must originate at cosmological distances. There has been no shortage of candidate sources. Fortunately, the measurements are now sufficiently definitive that many of these possible origins can be eliminated; they are briefly described here because, for a time, they were important to cosmologists and were motivating

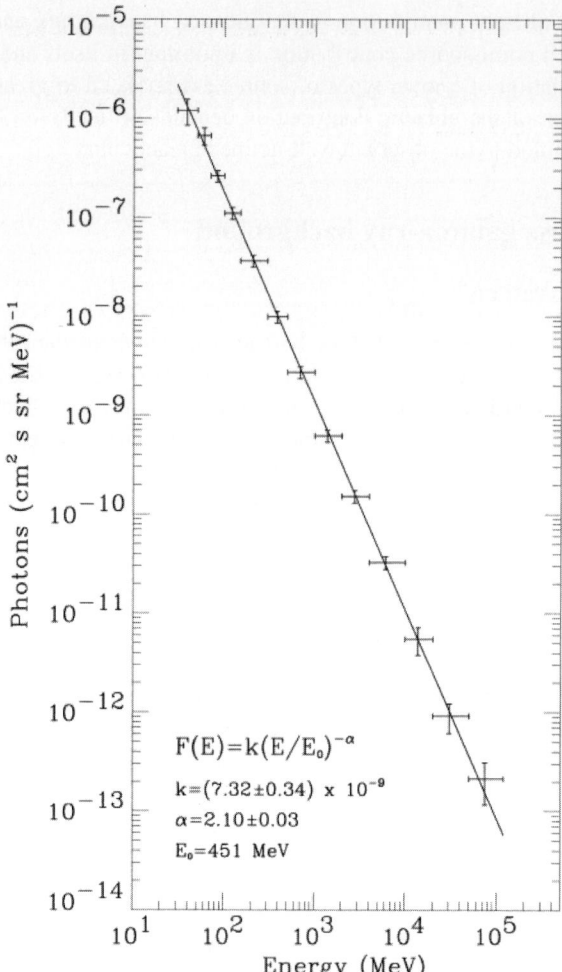

Figure 14.1. The extragalactic diffuse flux measured by EGRET at energies above 30 GeV [12]. (Reproduced with permission from the *Astrophysical Journal*.)

factors in the development of HE gamma-ray astronomy. It is convenient to divide these possibilities into those that are genuinely diffuse and those that are the sum total of discrete sources at large distances.

14.2.2.1 Truly diffuse

Before the Big Bang became the dominant concept in cosmology, there were strong advocates for a Steady State Universe in which baryons are spontaneously created in empty space. Equal amounts of baryons and anti-baryons are created.

This leads to annihilation with the production of gamma rays; the observed gamma-ray spectrum is the sum of all radiation out to $z = 100$. The red-shifted peak in the annihilation radiation is predicted to peak at low energies (≈ 1 MeV). The required creation rate leads to a flux that greatly exceeds the measured flux. In fact, one of the principal results from the Explorer XI experiment in 1964 was that the density ratio of anti-protons to protons was less than 10^{-6} [4]. Also there is now no observational evidence for what was once believed to be a bump in the diffuse spectrum at 1 MeV (see historical note: the 1 MeV lump).

Another possible scenario is that the universe was baryon-symmetric to begin with; this concept is attractive for a number of reasons and was enthusiastically pursued when it was believed that there was a bump at 1 MeV [13]. The most logical scale on which matter and anti-matter could be clumped would be in superclusters of galaxies. There is no evidence for excess gamma-ray emission at the boundaries of superclusters.

It was also suggested that the diffuse radiation might be the sum of the decay of primordial black holes (PBHs) created shortly after the Big Bang [8]. This attempt to unify quantum theory and general relativity led to the prediction that black holes would evaporate with the emission of HE and VHE gamma rays. Since the PBH lifetime is inversely proportional to the mass, eventually the evaporation is explosive with the exact mode of explosive decay depending on the assumed elementary particle model. The diffuse flux must be integrated over all distances and time and its spectrum has been calculated. However, the observed diffuse spectrum does not agree with the prediction, either in terms of spectral shape (break at 100 MeV) or overall intensity [5]. PBHs of mass 10^{15} decay in the present epoch and might be detectable as short bursts of gamma radiation. Experiments to search for the HE and VHE emission from nearby PBHs have also been unsuccessful [9, 6].

Other suggestions for the origin of the diffuse radiation include the decay of large black holes at extreme cosmological distance ($z \approx 100$) or the annihilation of supersymmetric particles. These, and other processes, predict features not confirmed by the EGRET observations.

14.2.2.2 Sum of discrete sources

It is natural that one should look to known discrete extragalactic sources as potential sources which, summed over all distances, might constitute the diffuse flux. Since normal galaxies make up the major component of the universe they are an obvious first choice. The Galaxy is the prototype of normal galaxies and we know its gamma-ray intensity and spectrum very well. Independent estimates of the intensity expected from the sum total of all normal galaxies agree that the contribution cannot be more than 10% of that observed above 100 MeV. In addition, the diffuse energy spectrum does not agree with that observed; the galactic spectrum is harder below 1 GeV and softer above it. Hence, while there is probably some component from normal galaxies, its contribution must be small.

Seyfert galaxies are known sources of hard x-rays but their spectra have not been shown to extend beyond 1 MeV. The radio galaxy, Cen A, has been detected at HE energies but such sources are unusual and do not constitute a major component.

The EGRET detection of 70 AGN, which are strong and variable sources of HE gamma rays, provides the best candidates for the diffuse background up to energies of 10 GeV (chapter 11). These are sources with hard spectra and they have been detected out to $z > 2$. The overall luminosity of an individual source depends on the degree of beaming but the total luminosity is independent of the beaming angle since, if the angle is small, there must be many more sources. It is striking that the average spectral index of all the observed blazars is $-(2.15 \pm 0.04)$ (chapter 11) which is in good agreement with the spectral index of the diffuse flux $-(2.10 \pm 0.03)$.

It is more difficult to estimate the total luminosity expected since evolutionary effects must be taken into account. The most straightforward way to do this is to assume that the gamma-ray luminosity of blazars evolves in the same way as the luminosity at shorter wavelengths, i.e. radio. However, the relationship between the radio and gamma-ray emission is not fully understood. The fact that the gamma-ray spectra of flat-spectrum radio sources and BL Lacs (which constitute the observed EGRET blazars) may be different complicates the issue. Another approach is to use the observed gamma-ray luminosities to derive an evolution function and, hence, predict the total intensity.

Many calculations agree that the observed intensity could be the sum of all blazars although some models find the contribution from unresolved blazars is only 25% of that required. With GLAST, it should be possible to detect many more AGN and to measure their energy spectra. It will then be possible to model the gamma-ray evolution function more accurately and thus confirm that the diffuse flux is truly the sum of all unresolved blazars. For the moment this is a good working hypothesis. It is somewhat of a disappointment since many high energy astrophysicists hoped for a more exotic solution.

14.3 Extragalactic background light

14.3.1 Stellar connection

Although we have previously stressed that high energy astrophysics in general, and HE/VHE gamma-ray astronomy in particular, have little in common with the thermal processes in the universe, there is one area of non-relativistic astrophysics where gamma-ray observations may make a major contribution. VHE, and to a lesser extent HE, gamma-ray measurements of AGN have an important overlap with thermal processes associated with stellar formation since the pair production of HE and VHE gamma rays on optical and infrared photons in extragalactic space limits their transmission. The thermal emission of these low energy photons is a tracer of star formation and, hence, the observations are of cosmological importance. The failure of VHE telescopes to detect many of the EGRET-detected AGN was originally explained [14] as due to the absorption of VHE gamma rays by pair production on infrared photons (see appendix). Although this explanation is no longer considered correct (it is more likely to be due to intrinsic differences in the gamma-ray emission spectra of FSRQs and BL Lacs (chapter 12)), it spurred interest in the potential of VHE observations to probe the density and spectrum of the extragalactic background light (EBL). Not much is known about the spectrum of the EBL at present nor how it developed over time. The VHE measurement is complicated by the fact that the gamma-ray *emission* spectra of AGN is not known.

Star formation is expected to be a major contributor to the EBL, with star formation contributing directly at short wavelengths (1–15 μm) and via dust absorption and re-emission at longer wavelengths (100–200 μm). The optical–infrared EBL background is, therefore, expected to have two broad peaks centered on these wavelengths (figure 14.2). The soft EBL background, $U_{EBL} = (h\nu)^2 n(h\nu)$ where $n(h\nu)$ is the density of photons of energy $h\nu$; this is conveniently quoted in units of nW m^{-2} sr^{-1}. U_{EBL} has values in the range 10–50 nW m^{-2} sr^{-1} near the two peaks.

14.3.2 Measurement of the soft EBL

Experiments that attempt to measure the EBL by directly detecting optical–infrared photons, such as the Diffuse Infrared Background Experiment (DIRBE) on the COsmic Background Explorer (COBE), are plagued by foreground sources of infrared radiation. Emitted and scattered light from interplanetary dust, emission from unresolved stellar components in the Galaxy, and dust emission from the interstellar medium are all significantly more intense than the EBL and must be carefully modelled and subtracted to derive estimates of the EBL. Currently, definitive EBL detections are available only at 140 and 240 μm. Detections at 2.2 μm and 3.5 μm and 400–1000 μm have also been reported.

Because VHE gamma rays are attenuated mostly by optical–infrared photons, measurements of the spectra of AGN provide an indirect means of investigating the EBL that is not affected by local sources of infrared radiation. Like direct measurements of the EBL, this technique has difficulties to overcome. For instance, it requires some knowledge of, or assumptions about, the intrinsic spectrum and flux normalization of the AGN. Also, the AGN themselves produce dense radiation fields which can absorb VHE gamma rays at the source and thereby mimic the effects of the intergalactic EBL attenuation.

Because the pair-production cross section is fairly strongly peaked around the photon energy (see appendix), the observed gamma-ray spectrum from a source, whose emission spectrum is a smooth power law, can be expected to have some of the features of the soft photon EBL in it. A sharp spectral feature in the soft photon EBL might result in a similar feature in the observed gamma-ray spectrum. The range of wavelengths 1–10 μm is of particular interest to the VHE

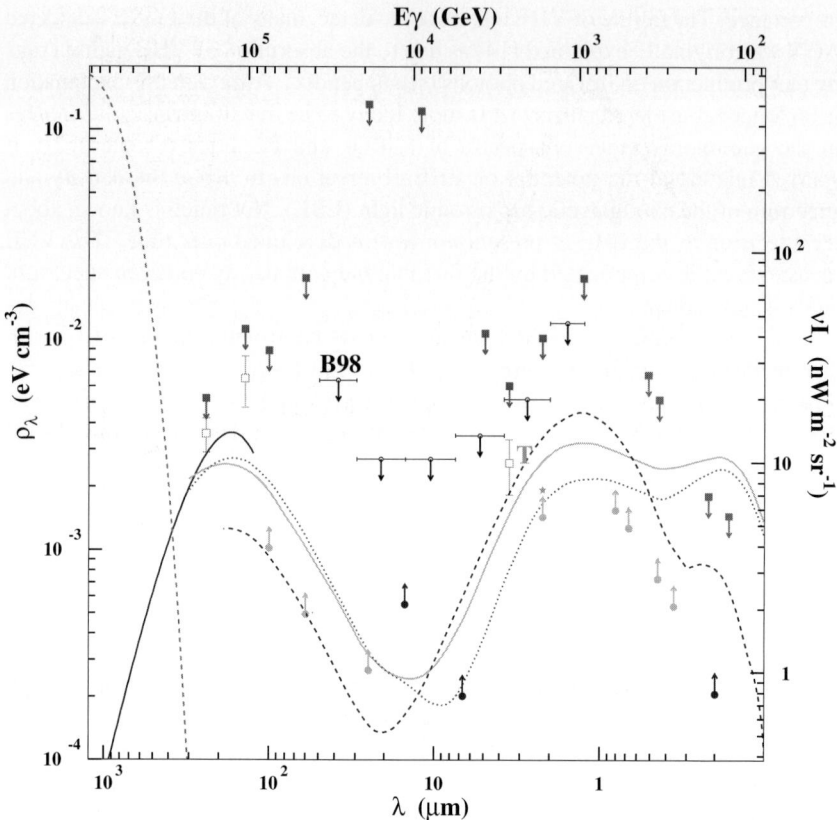

Figure 14.2. Measurements, limits and predictions of the diffuse intergalactic infrared background [15]. Upper limits derived from VHE gamma-ray spectra are indicated by the horizontal bars with arrows, marked as B98 [2]. Filled squares are upper limits from various experiments measuring the EBL directly. The open squares at 140 and 240 μm are detections from DIRBE. The open square marked 'T' indicates a tentative detection [3]. The filled circles are lower limits derived from galaxy counts. The full curve between 90 and 150 μm is a Far Infrared Absolute Spectrophotometer (FIRAS) detection. The broken line on the left indicates the 2.7 °K cosmic microwave background radiation. The three curves spanning most of the infrared wavelengths are different models of one study [10]. (Figure: V V Vassiliev.) (Reprinted from Catanese and Weekes 1999 *Publ. Astron. Soc. Pac.* **111** 1193. Copyright 1999 Astronomical Society of the Pacific; reproduced with permission of the editors.)

gamma-ray astronomer as it corresponds to the energy range currently accessible with ground-based telescopes. In practice, sharp features are not expected in thermal sources (which will be modified by cosmological effects anyway) so that

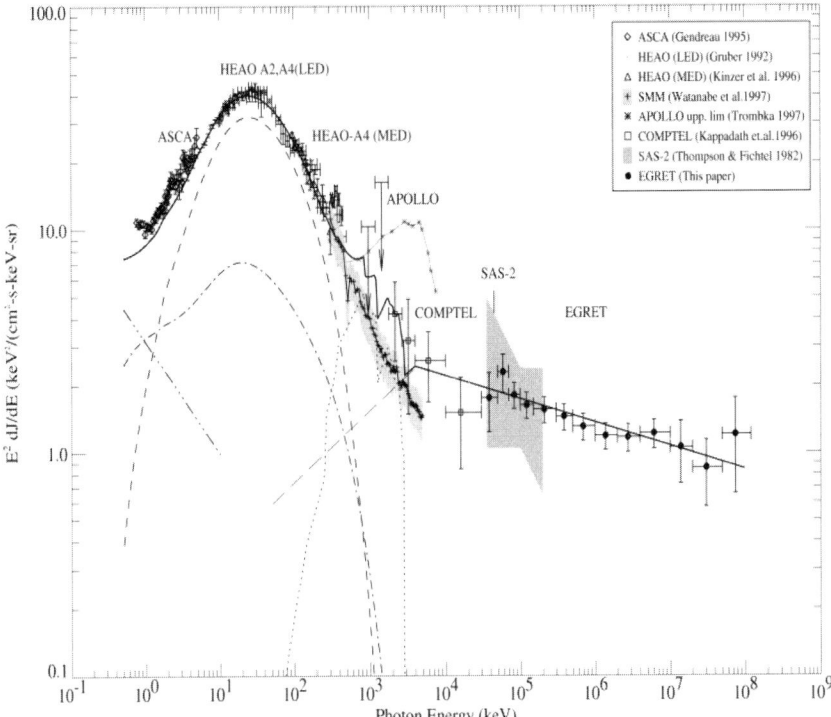

Figure 14.3. The diffuse background from x-ray to gamma-ray energies [11]. The good agreement between the SAS-2, COMPTEL and EGRET measurements are apparent. The early measurements agreed with the apparent detection of a 'bump' at 1 MeV (labelled 'Apollo' here). (Reprinted with permission from Sreekumar *et al* 1997 *AIP Conf. Proc. 410, Proc, 4th Compton Symp. 1997* ed Dermer and Kurfess, pp 344–58.)

the major effect of pair production on the EBL is a broad absorption over a range of energies, i.e. the effect will be an energy-independent absorption rather than a sharp spectral break [1].

14.3.3 VHE observations

Despite these difficulties, the accurate measurement of VHE spectra from the two confirmed VHE-emitting AGN, Mrk421 and Mrk501, has permitted some limits to be set on the density of the EBL over a wide range of wavelengths. These two sources are at practically the same distance and their spectra exhibit the same exponential cut-off in the range $E = 3$–6 TeV. This is consistent with EBL absorption but could also be a coincidence or a function of the source emission mechanism. The limits on the EBL density have been derived from two approaches: (1) assuming a limit to the hardness of the intrinsic spectrum of the AGN and deriving limits which assume very little about the EBL spectrum [2]; and (2) assuming some shape for the EBL spectrum, based on theoretical or phenomenological modelling of the EBL, and adjusting the normalization of the EBL density to match the measured VHE spectra [7]. The latter can be more stringent but is necessarily more model-dependent. The limits from these indirect methods and from the direct measurements of EBL photons are summarized in figure 14.2. At some wavelengths, the TeV limits represent a 50-fold improvement over the limits from DIRBE. These limits are currently well above the predicted density for the EBL from normal galaxy formation [10] but they have provided constraints on a variety of more exotic mechanisms for sources of the EBL. They also show that EBL attenuation alone cannot explain the lack of detection of EGRET sources with nearby red-shifts at VHE energies. The optical depth for pair production does not reach 1 for the stringent VHE limits [2] until beyond a redshift of $z = 0.1$.

Historical note: the 1 MeV bump

Scientific research does not always proceed along a straight road but often stumbles into interesting pot holes along the way. Such was the case with the diffuse background. Although it was always recognized that the region around 1 MeV would be particularly difficult because of the instrumental background from radioactivity in the detector and its surrounds, the early results from balloon experiments and from measurements made by detectors on the Apollo missions to the moon agreed that there was an excess in the region near 1 MeV although there was disagreement about its absolute value. The balloon experiments were particularly difficult because, in addition to radioactivity in the detector components, there was secondary radiation from cosmic rays in the atmosphere; it was necessary to make measurements at different depths of atmosphere and then to attempt to extrapolate to zero atmosphere. The Apollo spacecraft were large and offered the possibility of much secondary radiation from the spacecraft itself. Some measure of this could be determined by operating the detector on a long arm at different distances from the spacecraft. The possibility of a distinct feature in the diffuse background at 1 MeV led to much theoretical speculation with diverse cosmological explanations proposed. These engendered a lively debate about the possibility that they might be due to antimatter–matter annihilation. Finally, the subject was put to rest when definitive measurements by COMPTEL and the Solar Maximum Mission (SMM) became available. These agree and join smoothly with the x-ray spectrum (figure 14.3): no new diffuse component is required.

Recent, but less accurate, measurements of the VHE spectra of H1426+428 and 1ES1959+650 allow these conclusions to be extended since these two sources are at larger red-shifts but otherwise seem very similar to Mrk421 and Mrk501. The measured VHE energy spectrum of H1426+428 is steep and may indicate that the emitted flux above 1 TeV is some 100 hundred times greater than the measured flux. Hence, the VHE gamma-ray luminosity is ten times greater than the observed contemporaneous x-ray luminosity [1]. This is not expected in the simple Synchrotron-Self Compton models.

The new generation of VHE telescopes will lead to the detection of more AGN at a variety of red-shifts; with improvements in our understanding of the emission and absorption processes in AGN, these VHE measurements have the potential to set very restrictive limits on the EBL density and, perhaps, eventually detect it.

Measurements of the EBL have the potential to provide a wealth of information about several other important topics in astrophysics. In addition to putting some limits on the history of the formation of stars and galaxies [3], they can limit other, more exotic processes, such as pre-galactic star formation and some dark matter candidates, which might contribute distinctive features to the EBL [2].

References

[1] Aharonian F A 2001 *Proc. 27th ICRC (Hamburg, August 2001)* ed K H Kambert, G Heinzelmann and C Spiering (University of Hamburg) p 250
[2] Biller S D *et al* 1998 *Phys. Rev. Lett.* **80** 2992
[3] Dwek E *et al* 1998 *Astrophys. J.* **508** 106
[4] Fazio G G 1967 *Annu. Rev. Astron. Astrophys.* **5** 481
[5] Fichtel C E and Trombka J I 1997 *Gamma Ray Astrophysics (NASA Ref. Publ. 1386)* p 219
[6] Fichtel C E *et al* 1994 *Astrophys. J.* **434** 557
[7] de Jager O C, Stecker F W and Salamon M H 1994 *Nature* **369** 294
[8] Page D N and Hawking S W 1976 *Astrophys. J.* **206** 1
[9] Porter N A and Weekes T C 1978 *Mon. Not. R. Astron. Soc.* **183** 285
[10] Primack J R, Bullock J S, Somerville R S and MacMinn D 1999 *TeV Astrophysics of Extragalactic Sources (Astropart. Phys. 11)* ed M Catanese and T C Weekes (Amsterdam: North-Holland) p 93
[11] Sreekumar P, Stecker F W and Kappadath S C 1997 *Proc. 4th Compton Symposium (AIP Conf. Proc. 410)* ed C D Dermer, M S Strickman and J D Kurfess (New York: AIP) p 344
[12] Sreekumar P *et al* 1998 *Astrophys. J.* **494** 523
[13] Stecker F W 1989 *Proc. Gamma Ray Observatory Science Workshop* ed W N Johnson (Goddard Space Flight Center: NASA) pp 4–73
[14] Stecker F W, de Jager O C and Salamon M 1992 *Astrophys. J. Lett.* **390** L49
[15] Vassiliev V V 2001 *Astropart. Phys.* **12** 217

Appendix

Radiation and absorption processes

A.1 Introduction

There are a vast range of radiation processes covered by very-high energy (VHE) gamma-ray astronomy phenomenology and it would be inappropriate to try to cover them here in any depth. There are a few processes that are of particular interest to the high energy gamma-ray astronomer and these will be briefly described. These are Compton scattering (fundamental to Compton telescopes and, in its inverse form, one of the main production mechanisms in sources), pair production with matter (the key process for detecting gamma rays in space telescopes at energies above 10 MeV), electron bremsstrahlung (an important mechanism for the production of gamma rays in the Galaxy), pion production (the major mechanism by which gamma rays are produced by hadrons), photon–photon pair production (the only important absorption mechanism for gamma rays at these energies), synchrotron radiation (the principal emission mechanism from high energy particles in astrophysical situations), and Cherenkov radiation (the essential mechanism for the detection of VHE gamma rays by ground-based telescopes).

In the energy region above 1 keV, there are basically three processes by which the gamma rays can interact with the matter: these are the photoelectric effect, Compton scattering, and pair production. The relative cross sections or, more practically, the mass absorption coefficients peak in different energy ranges (figure A.1). This has the same functional form for all materials, although the actual values and relative strengths of the three processes differ. As the photoelectric effect, the interaction of the gamma ray with bound electrons in atoms, is only important at low energies (<1 MeV), we shall not consider it here.

A.2 Compton scattering

The scattering of a photon off an unbound electron is known as Compton scattering. This is important both in the production and detection of gamma

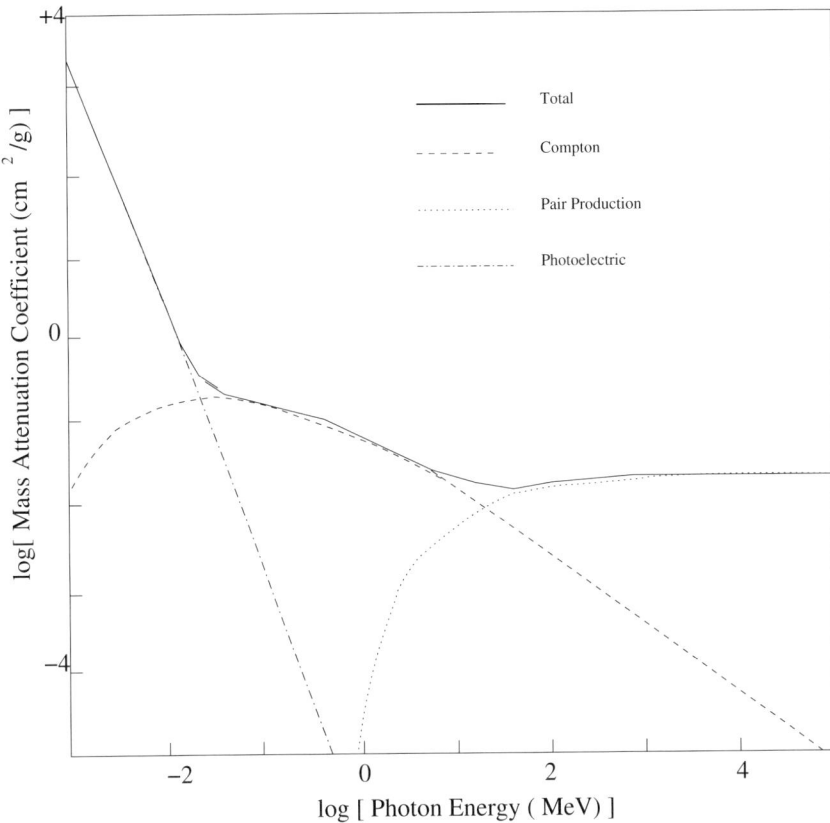

Figure A.1. The mass attenuation coefficient for various interactions in a plastic scintillator.

rays. It is easily understood when the photon is treated as a particle and its wave properties ignored. It is the dominant interaction of a gamma ray in the energy region from a few hundred keV to 10 MeV in most materials and is the basis of the Compton telescope. Hence, it is most important when the photon energy is of order or greater than the rest mass of the electron. Initially it is assumed that the electron is at rest or that we are working in the rest frame of the electron.

The inverse Compton process, the collision of a high energy electron with a low energy photon, is found to be very important in astrophysical systems. It is physically the same as Compton scattering and can be described in terms of a coordinate transformation to the rest frame of the electron. To the observer in that frame, the stationary electron scatters an energetic photon. Since the scattered photon acquires considerable energy it can be envisaged as a form of photon 'energy boosting'. It is the dominant mechanism by which VHE gamma rays are produced by electrons in astrophysical sources such as plerions and AGN.

The scattering can be characterized by a set of equations which arise from simple physical considerations of conservation of energy and momentum [2]. The geometry and physical parameters are defined in figure A.2(a). Also

$$E = m_e c^2 [1 - v^2/c^2]^{-1/2} = \gamma m_e c^2$$
$$\alpha = h\nu/m_e c^2$$

and

$$r_0 = (e^2/m_e c^2) = \text{the classical electron radius.}$$

- Conservation of energy:

$$m_e c^2 + h\nu = E + h\nu'$$

where E = the total energy of scattered electron.
- Conservation of momentum:

$$h\nu/c = (h\nu'/c)\cos\theta + \gamma m_e v \cos\phi.$$

- Conservation of momentum:

$$0 = (h\nu'/c)\sin\theta - \gamma m_e v \sin\phi.$$

These equations can be solved to give the simple relationships:

$$c/\nu - c/\nu' = (h/m_e c)(1 - \cos\theta)$$
$$h\nu' = (m_e c^2)/[1 - \cos\theta) + (1/\alpha)].$$

If $\theta = 180°$, i.e. backward scattering and α is large, the energy of the scattered photon is about 50% of $m_e c^2$ (or 0.255 MeV). If $\theta = 90°$, the photon energy is nearly 0.51 MeV. It should be noted that in Compton scattering, low energy photons suffer only a small energy change whereas at higher energies the change is proportionally much larger. For large α, the scattered gamma radiation is predominantly in the forward direction.

To evaluate the cross section, it is necessary to treat the process rigorously with quantum mechanics [2]. This yields the Klein–Nishina cross section:

$$\sigma_{KN} = \pi r_0^2 [(1/\alpha^3) \ln(1 + 2\alpha) + 2(1 + \alpha)(2\alpha^2 - 2\alpha - 1)$$
$$\times \{(\alpha^2(1 + 2\alpha)^2) + 8\alpha^2/3(1 + 2\alpha)^3\}^{-1}] \quad \text{cm}^2/\text{electron}$$

or, for small values of α,

$$\sigma_{KN} = (8\pi/3)r_0^2)(1 - 3\alpha + 9.4\alpha^2 - 28.0\alpha^3 + \cdots) \quad \text{cm}^2/\text{electron.}$$

For very small values of α, this reduces to the Thompson cross section:

$$\sigma_T = (8\pi/3)r_0^2 = 0.665 \times 10^{-24} \text{ cm}^2 = 0.665 \text{ barns.}$$

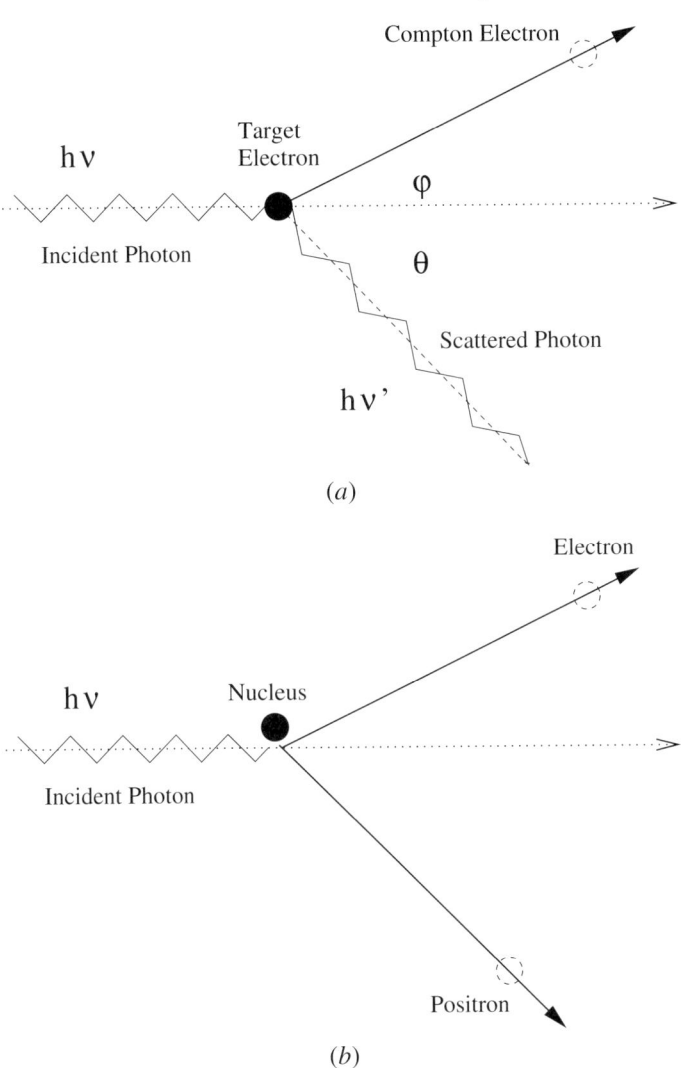

Figure A.2. (*a*) Compton scattering; (*b*) pair production; (*c*) electron bremsstrahlung; (*d*) pion production.

The Compton scattering cross section decreases only slowly with energy.

Detectors which depend on Compton scattering are the most complicated and the most difficult to interpret. The gamma ray may undergo one or more Compton scatterings, losing energy to electrons in each case, until eventually it may undergo a photoelectric reaction. In each scattering, the electron will take up some of the energy as kinetic energy and the gamma ray will change direction.

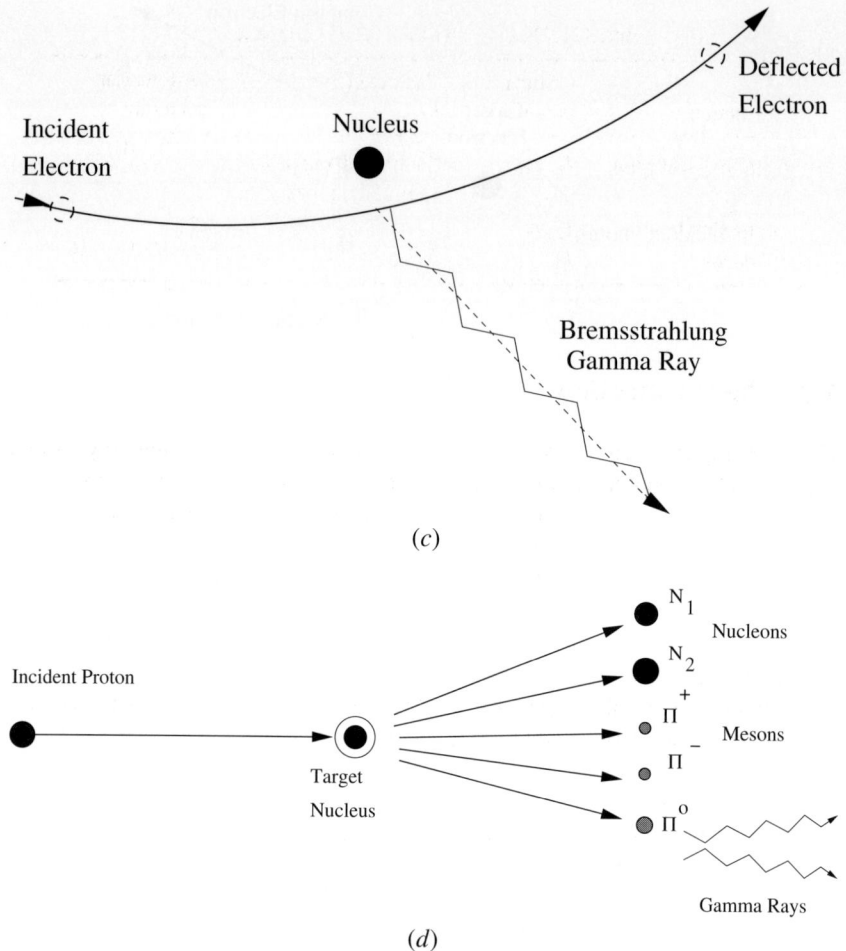

Figure A.2. (Continued.)

Because of this complexity, observational gamma-ray astronomy in the energy range from 100 keV to 30 MeV has been the slowest discipline to develop.

If inverse Compton scattering is considered from relativistic electrons which have a power-law distribution of the form $I_e(E_e) = K_e E^{-\Gamma_e}$, and the soft photons have a density ρ_{ph}, then the resulting gamma rays will have characteristic energies of $(\gamma_e)^2 h\nu$ (Thompson) or $\gamma_e h\nu$ (Klein–Nishina) where γ_e is the electron Lorentz factor and $h\nu$, the soft photon energy [3, 4]. The resulting differential spectrum will be proportional to $E_\gamma^{-(\Gamma_e+1)/2}$.

Table A.1. Production of HE/VHE gamma rays.

Interaction	Critical parameters	Characteristic E_γ	Differential spectrum
Compton scattering	I_e, ρ_{ph}	$(\gamma_e)^2 h\nu$ (Thompson) $\gamma_e h\nu$ (Klein–Nishina)	$E_\gamma^{-(\Gamma_e+1)/2}$
Bremsstrahlung	I_e, ρ_g	E_e	$E_\gamma^{-\Gamma_e}$
π^0 decay	I_p	70 MeV	$E^{-4/3(\Gamma_p-1/2)}$

A.3 Pair production

The most important energy loss mechanism for HE and VHE gamma rays is pair production with matter. In this interaction, the incident gamma ray is completely annihilated with its energy transferred to an electron pair which is created, i.e. $h\nu \to e^+ + e^-$ (figure A.2(b)).

The interaction takes place in the electric field of a nucleus which takes up some of the momentum. Obviously the threshold for the interaction must be >1.02 MeV (i.e. $2m_e$) [5]. The interaction can also occur in the field of an electron but the cross section is much less and the threshold is higher. The energy of the gamma ray is taken up by the electron pair as rest mass and kinetic energy. The pair are strongly beamed forward and the trajectory of the gamma ray can be inferred from their trajectories. Note that the kinetic energies are not equally shared and the initial trajectory is not necessarily the mean of their emission angles. The positron will generally annihilate later with an electron to produce two gamma rays, which can Compton scatter or suffer photoelectric absorption. These secondary products at moderate energies can be totally absorbed in the detector; by measuring their energy and adding the rest mass of the pair, the energy of the incident gamma ray is estimated.

The cross section for pair production rises rapidly and becomes dominant over the alternative processes above energies of about 30 MeV, after which it rises slowly to an asymptotic value (figure A.1). This value is given by

$$\sigma_{pp} = \sigma_0 Z^2 [(28/9) \ln(183/Z^{1/3}) - 2/27] \quad \text{cm}^2 \text{ atom}^{-1}$$

where $\sigma_0 = (1/137)(e^4/(m_e^2 c^4)) = 5.8 \times 10^{-28}$ cm^2/nucleus $= 0.58$ millibarn. At very high energies the cross section is energy independent.

Pair production is the key process in HE space telescopes; it also plays a vital role in the development of the atmospheric electromagnetic cascades that make VHE gamma-ray astronomy possible. The mean distance that a gamma ray travels before it undergoes pair production is given by

$$\lambda_{pp} = 1/(N\sigma_{pp})$$

where N is the number of target nuclei per unit volume. The mean free path for pair production is related to the radiation length, X_0:

$$\lambda_{pp} = 9/7 X_0.$$

The physics of the pair-production interaction is vital in the design of HE pair-production telescopes. The tracks of two ionizing particles originating in a common point of origin is very characteristic and easy to recognize. A critical property for practical detectors is the degree to which the electron pair maps the projected trajectory of the gamma ray. The root-mean-square angle between the trajectory of the secondary electron of energy E_e and that of the primary gamma ray of energy E_γ is about 4° at $E_\gamma = 30$ MeV, 1.5° at $E_\gamma = 100$ MeV, and 0.2° at $E_\gamma = 1$ GeV. The value of observations at high energies is thus apparent.

Unfortunately, it is not possible to measure the electron trajectory precisely as it will inevitably undergo Coulomb scattering as it passes through the material of the detector.

Measuring the energy of the gamma ray is essentially measuring the energy of the electron pair and their secondary products. This requires that the detector have sufficient mass to absorb all these products. In practice, this calls for a calorimeter with as much absorber in the payload as the spacecraft can carry.

A.4 Electron bremsstrahlung

When an incident charged particle is deflected in the electric field of a nucleus, it emits electromagnetic radiation whose amplitude is proportional to the acceleration causing the deflection. In the classical case, the acceleration, produced by a nucleus of charge Ze on a particle of charge e and mass m is proportional to Ze^2/m. This acceleration, which is actually a deceleration of the incident particle, is called bremsstrahlung (German for braking radiation).

In the astrophysical situation, this process is most important for relativistic electrons in the presence of atomic or molecular material; these will be deflected and emit gamma rays as bremsstrahlung radiation [8]. The process may be particularly important for cosmic electrons in the SNR and in the interstellar medium.

The complete quantum mechanical treatment of electron bremsstrahlung by an atom is complex because of the effects of screening by the atomic electrons and the finite nuclear radius [2]. Both classical and quantum mechanical treatments give *average* cross sections for many bremsstrahlung interactions of the same order: $\sigma_b \approx \sigma_0 = 4(1/137)(e^2/m_0 c^2)^2 Z^2 = 0.58$ millibarn/nucleus. This is remarkable since in the classical case the process is continuous with multiple small emissions as the electron traverses the material; in the quantum mechanical case, the probability of an individual emission is small but, when it does occur, the emitted quantum has energy comparable to the incident electron. The following approximate expressions for σ_b have been derived for different energy ranges of T, the kinetic energy of the electron:

- Non-relativistic case; $T < m_e c^2$:

$$\sigma_b = (16/3)\sigma_0 Z^2 \quad \text{cm}^2/\text{nucleus}.$$

- Mildly relativistic case; $T \approx m_e c^2$: No analytical expression is available.
- Highly relativistic; $T > m_e c^2$:

$$\sigma_b = 4[\ln(2(T + m_e c^2)/m_e c^2) - 1/3]\sigma_0 Z^2 \quad \text{cm}^2/\text{nucleus (averaged)}.$$

- Extreme relativistic; $T > 137 m_e c^2 Z^{-1/3}$:

$$\sigma_b = 4[\ln(183 Z^{-1/3})]\sigma_0 Z^2 \quad \text{cm}^2/\text{nucleus (averaged)}.$$

In the astrophysical case, one is interested in gamma-ray production from a spectrum of cosmic electrons in a gas. If the gas density is ρ_g, then the production of gamma rays depends on ρ_g and the electron energy distribution. The gamma rays that result from bremsstrahlung have energies of the same order as the incident electron so that if the electron population is characterized by a power law with spectral index, Γ_e, the resulting gamma-ray spectrum has an index Γ_γ and $\Gamma_e \approx \Gamma_\gamma$ (table A.1).

A.5 Pion production

One of the most common interactions of cosmic ray protons in astrophysics is collision with stationary hydrogen gas, producing excited states that lead to the emission of π mesons. The threshold kinetic energy of the incident proton is 290 MeV. The most common interaction has the form:

$$p + p \rightarrow N + N + n_1(\pi^+ + \pi^-) + n_2(\pi^0)$$

where N is a proton or neutron and n_1 and n_2 are integers (figure A.2(d)). Below 1 GeV, $n_1 = n_2 = 1$. At high energies the cross section for π production is constant and equal to 27 millibarn. The π^0's decay into two gamma rays with a half life of 10^{-16} s. In the rest frame of the π_0, each gamma ray has an energy of $m_\pi \approx 70$ MeV. If the cosmic rays have a power-law spectral distribution with index Γ_p, then at high energies the gamma-ray spectral distribution will also be a power-law with $\Gamma_\gamma = 4/3(\Gamma_p - 1/2)$ (table A.1). As the energy decreases, the spectrum turns over with a peak at 70 MeV. It is this peak that is the characteristic feature of the p–p interaction and the signature of hadrons as the progenitors in cosmic gamma-ray sources.

Strictly speaking, the decay of the excited states of the proton into K mesons and hyperons should also be taken into account but these are generally ignored as they are infrequent.

A.6 Gamma-ray absorption

Gamma rays are notorious for their penetrating power. However, there are certain conditions under which absorption must be taken into account.

A.6.1 Pair production on matter

That gamma rays interact with matter is obvious from the fact that they cannot penetrate the earth's atmosphere and can interact in space gamma-ray telescopes. These processes have already been described. At high energies the most important process is pair production in the presence of hadronic or leptonic matter. The radiation length is approximately 38 g cm^{-2} and the cross section approximately 10^{-26} cm^2 or 0.01 barns. The typical density of interstellar space is about 1 atom cm^{-3}; in intergalactic space it is more like 10^{-5} atoms cm^{-3}. Typical interstellar distances are 10 000 light-years (10^{22} cm) and intergalactic distances 100 million light-years (10^{26} cm). With atoms of mass approximately 10^{-24} g, the amount of matter encountered in travelling from sources at these distances is much less than a radiation length so that the absorption of the gamma-ray beam by matter will be negligible.

However close to, or in, a source, where matter densities may be much higher, this is a process that must be taken into account.

A.6.2 Photon–photon pair production

This is a process that is almost unique to the astrophysical situation since it requires unusual combinations of high energy photons and a high density of lower-energy photons. Gamma rays are absorbed by photon–photon pair production ($\gamma + \gamma \rightarrow e^+ + e^-$) on background photon fields if the center-of-mass energy of the photon–photon system exceeds twice the rest energy of the electron squared. The cross section for this process peaks when

$$E_\gamma h\nu(1 - \cos\theta) \sim 2(m_e c^2)^2 = 0.52(\text{MeV})^2 \qquad (A.1)$$

where E_γ is the energy of the γ-ray, $h\nu$ is the energy of the low energy photon, θ is the collision angle between the trajectories of the two photons, m_e is the mass of the electron, and c is the speed of light in vacuum. Thus, for photons of energy near 100 MeV, head-on collisions with x-ray photons of ∼5 keV have the highest cross section. Dense fields of x-ray photons may be encountered in the immediate vicinity of a 100 MeV source, e.g. in the accretion disks surrounding AGN, where this effect must be taken into account. However, in interstellar and intergalactic space, the ambient x-ray background is small and photon–photon pair production is negligible for HE gamma rays except at extreme cosmological distances.

The effect is more important for VHE gamma-ray astronomy since here photons of energy 1 TeV have the maximum cross section for head-on collisions with near infrared photons of energy 0.5 eV ($\lambda \sim 2\ \mu$m). There is no shortage of stellar and dust sources of this radiation in the K-band so that even within the Galaxy absorption may not be negligible. The absorption is particularly important for extragalactic sources where the presence of extragalactic background light (EBL) limits the distance to which VHE gamma-ray telescopes can detect sources (chapter 14).

For interactions from a source at a distance corresponding to a redshift, z, equation (A.1) becomes:

$$E_\gamma(z)(1+z)\epsilon(1+z)x \approx 2(m_e c^2)^2 = 0.52 \text{ (MeV)}^2 \quad \text{where } x = (1 - \cos\theta).$$

It can be shown that the pair creation cross-section is given by

$$\sigma[E(z), \epsilon(z), x] = 1.25 \times 10^{-25}(1-\beta^2)[2\beta(\beta^2 - 2) \\ + (3-\beta^4)\ln(1+\beta)/(1-\beta)] \quad \text{cm}^2$$

where $\beta = 1 - 2(m_e c^2)^2/[E\epsilon x(1+z)^2]^{1/2}$.

The attenuation over a distance, d, is characterized by the the optical depth $\tau(E) = d/L(E)$ where $L(E)$ is the mean free path. A convenient approximation for the optical depth is

$$\tau(E) \approx U_{\text{EBL}}(\nu Ez/H)$$

where $U_{\text{EBL}} = (h\nu)^2 n(h\nu)$ in units of 10 nW m^{-2} m^{-2} sr^{-1}, E is the gamma-ray energy in TeV, z = is the redshift in units of 0.1, and H is the Hubble Constant in units of 60 km s^{-1} Mpc^{-1} [1]. These values are appropriate to the nearby blazars detected at VHE energies and estimated energy density of the EBL at infrared wavelengths.

A.7 Synchrotron radiation

The discovery of polarized radio emission from supernova remnants and radio galaxies was explained by Russian physicists in the post Second World War era as examples of synchrotron radiation in cosmic settings. It is now universally recognized that the same radiation process that is observed from relativistic particles in the strong magnetic fields of manmade particle accelerators is at play in the emission from ultra-relativistic particles in the generally much weaker magnetic fields in these cosmic sources. A non-relativistic electron moving through a homogeneous magnetic field follows a helical path around the lines of force. The motion consists of two components: one is parallel to the lines of force; and the other is rotation about them at the angular frequency of Larmor precession:

$$\omega_L = eH/mc$$

where H is the intensity of field normal to the velocity vector of the electron. The electron radiates like a dipole with frequency ω_L [8].

At relativistic energies, the radiation is more complex since the radiation is beamed into a cone of angle $\theta \approx m_e c^2/E$ (figure A.3). An observer located in the

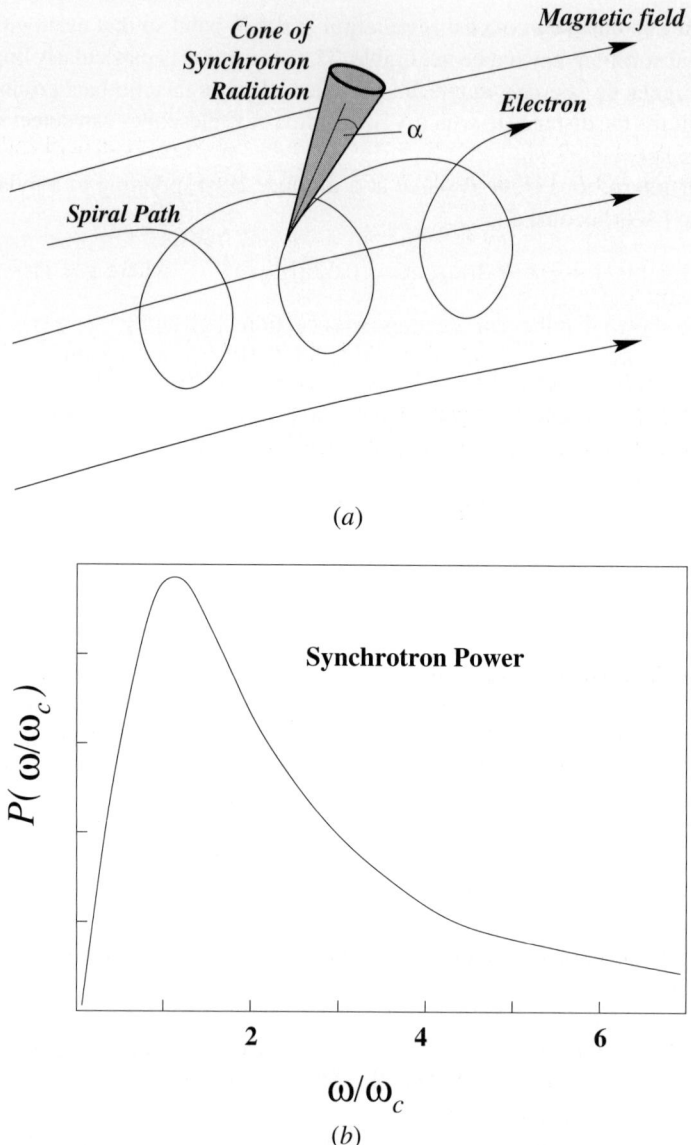

Figure A.3. (*a*) The geometry of synchrotron emission from a particle in a magnetic field; (*b*) the power distribution as a function of critical frequency.

orbital plane of the electron will only detect radiation when the cone is pointed in that direction. Instead of occurring at a single frequency, the radiation now occurs as a continuum spectrum distributed as shown in figure A.3 about ω_c, the critical frequency at which the maximum power is emitted.

$$\omega_c = (3/2)(eH/mc)\gamma^2 \sin\phi$$

where ϕ is the pitch angle between the direction of the magnetic field and that of the electron. For H in microgauss and E in GeV, this gives

$$\omega_c \approx 100\, H E^2 \sin\phi \quad \text{Mhz}.$$

The energy loss is given by:

$$-dE/dx = 1/c\, dE/dt = (2e^4/3m^2c^4)\gamma^2 H^2 \quad \text{erg cm}^{-1}$$

where E is in ergs and H in gauss [12].

The power distribution above and below ω_c are given by below ω_c:

$$P(\omega/\omega_c) = 0.256(\omega/\omega_c)^{1/3}$$

above ω_c:

$$P(\omega/\omega_c) = 1/16(\pi\omega/\omega_c)^{1/2} \exp[-2\omega/3\omega_c].$$

A.8 Cherenkov radiation

Cherenkov radiation occurs when a particle travels through a dielectric medium with a velocity that exceeds the velocity of light in that medium. It is relevant to relativistic particles, the radiation occurs over a broad band, and there is a threshold velocity for emission ($v/c > 1/n$ where v is the velocity of particle, c the velocity of light, and n, the refractive index). The radiation is emitted at an angle that depends on the refractive index and is beamed in the forward direction (figure A.4). The most comprehensive treatment of the topic can be found in Jelley's classic book on the subject [9] which, although published in 1958, is still the best reference.

When a charged particle passes through a dielectric medium, it interacts electrically with the molecules in its immediate vicinity. It disturbs the neutrality of the molecules inducing polarization that turns on and off as the particle passes and causes the molecule to radiate. If the particle is slow moving, the disturbance is symmetrical around and along the particle trajectory so that there is no residual electric field and, hence, no detectable radiation [10]. This is illustrated in figure A.5(a) where the particle is an electron and the medium is a solid. It is easiest to consider these molecules to be closely spaced as in a solid or liquid (where the effect was first seen) although the same principles apply to a gas, e.g. the earth's atmosphere.

If the particle is moving at relativistic velocity, the situation is quite different. In this case the particle velocity, v, exceeds the velocity of light in the medium, c/n, where n is the refractive index (figure A.5(b)). In the radial direction,

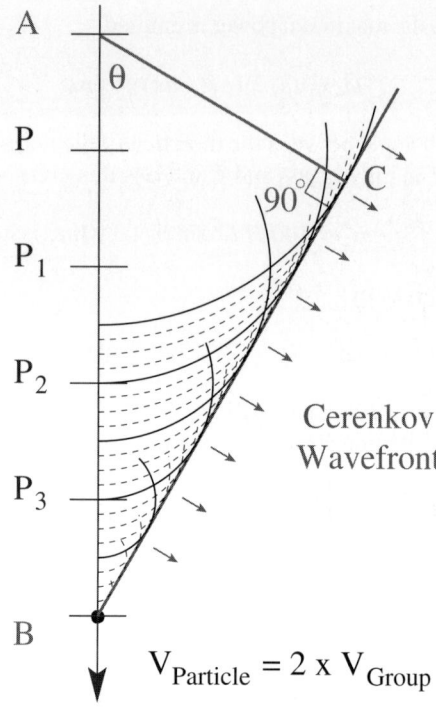

Figure A.4. The coherence condition for Cherenkov radiation in a solid material with large index of refraction. (Figure: D Horan.)

symmetry is still preserved but along the trajectory there will be a resultant dipole field in the medium which can produce detectable effects. As the particle traverses the dielectric, each finite element radiates a brief electromagnetic pulse.

Although the wavelets in the pulse will interfere destructively in general, in the forward direction the wavefront from each element of track will interfere constructively as seen in the Huygens wavelet reconstruction in figure A.4. From the figure it is seen that the angle θ is determined from the relative values of v and n, according to: $\cos\theta = (n/c)/v$. This is the fundamental Cherenkov equation. Clearly, there is a threshold velocity where $v/c = 1/n$, a maximum Cherenkov angle where $v = c$ and the radiation will only occur where $n > 1$ which covers the optical region of the spectrum for most materials. It is these properties of well-defined emission angle and threshold velocity that make Cherenkov radiation detectors so useful in particle physics.

There is an analogy between Cherenkov radiation for light and supersonic shocks for sound. Just as an object will only produce a sonic boom when it

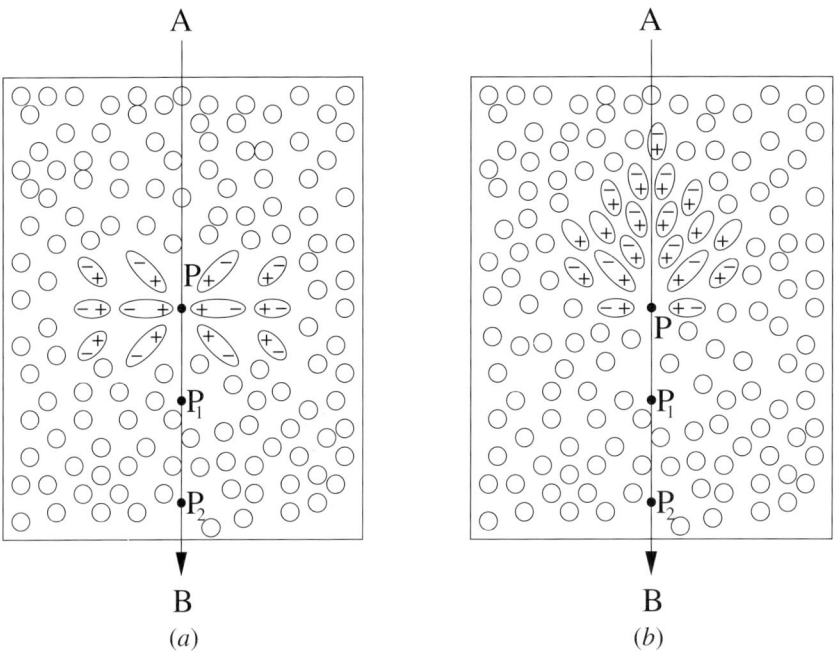

Figure A.5. The local polarization produced in a medium during the passage of a fast particle. (Figure: D Horan.)

exceeds the velocity of sound in the medium, a particle must exceed the velocity of light in the medium.

The rigorous theory of Cherenkov radiation (more correctly called Cherenkov–Vavilov radiation after its co-discoverers) was developed by Frank and Tamm from which the basic formulae are derived [6].

From this theory the emission formula is derived:

$$dE/dt = (e^2/c^2) \int \sin^2\theta \, \omega \, d\omega.$$
$$\text{I}\text{II}\text{III}$$

The basic components of this equation were expounded by Jelley [11]. The mechanism is quite different from other, more familiar, radiation mechanisms such as synchrotron radiation or bremsstrahlung. The three factors in the Cherenkov formula for radiation yield can be understood by analogy with the single elementary classical dipole.

The intensity of radiation from a dipole is given by $(i^2 Z)$ where i is the current and Z, the radiation resistance. This can be taken as represented by the (charge)^2 term in the Cherenkov formula (I).

The angular distribution can be understood by realizing that the short element of track behaves like a simple dipole. A stationary dipole radiates with an angular distribution of $\sin^2 \theta$ where θ is the angle made with the trajectory and the direction of the observer (II).

Consider how the net polarization is seen from some arbitrary point to the side of the particle trajectory. As the particle moves, the direction of the observed polarization changes. If the radial and axial components are considered as a function of time, the observer sees no residual radial component because of the axial symmetry; however, the axial component will appear as a double δ-function. The Fourier transform of this function gives the spectral distribution of Cherenkov radiation proportional to $\omega \, d\omega$ (III).

Note the following properties:

- The process is a macroscopic one in which the medium as a whole is involved.
- The medium is what produces the radiation, not the particle itself.
- Quantum effects are unimportant because the energy of the emitted photons is very small compared to that of the particle.

In water, where $n = 1.33$, θ_{max} is of order $41°$ and for electrons, the threshold energy, $E_t = 260$ keV and the Cherenkov photon yield is 2500 photons m^{-1}. In the atmosphere at ground level, $n = 1.00029$ and θ_{max} is $1.3°$, E_t for electrons is 21 MeV and for muons, 4 GeV. The light yield in the visible range is about 30 photons m^{-1} or 10^4 photons per radiation length.

Historical note: distance limit

The importance of photon–photon interactions for gamma-ray measurements was first pointed out by Nikishov [13] in 1962 who calculated its effect for TeV photons. Using the best available estimates for the density of starlight at the time (about 0.1 eV cm^{-3}), he initially found a value for the absorption coefficient, $k = 7 \times 10^{-27}$ cm^{-1} at 1 TeV. This large attenuation had a chilling effect on prospective projects in TeV gamma-ray astronomy then under consideration. A re-evaluation by Gould and Schreder [7] showed that the starlight density had been over-estimated by two to three orders of magnitude (figure A.6.); hence, the effect is not critical for galactic sources or even nearby extragalactic sources. However, the discovery of the cosmic blackbody microwave background at $2.7\,°$K led to the prediction of very strong absorption of gamma rays in the 10^{14}–10^{16} eV bands and the virtual confinement of gamma-ray studies at these energies to galactic sources. This led to a virtual moratorium on the construction of new air shower arrays whose sensitivity was in that energy range. It was not until the apparent detection of Cygnus X-3 in 1983 that interest in building arrays sensitive to 100 TeV gamma rays was revived. Similar strong absorption is predicted for gamma rays above 10^{18} eV by extragalactic radio photons (figure A.6).

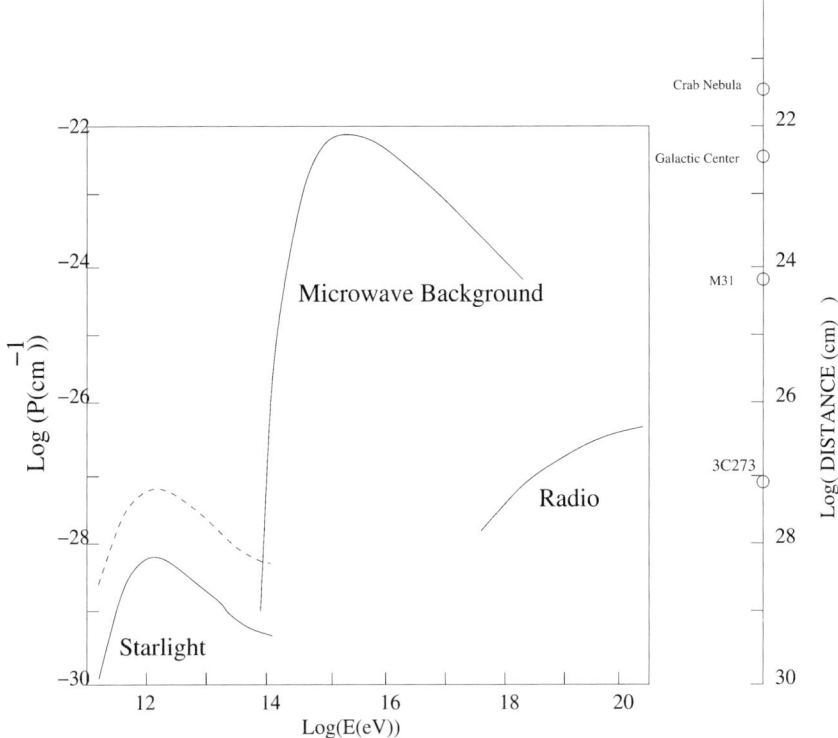

Figure A.6. The absorption coefficient for pair production on the diffuse background radiation as a function of incident photon energy. For comparison the distance to representative objects is shown on the right.

The amount of energy that goes into this process is negligible. The size of the energy exchange between the relativistic particle and an individual molecule is of order 4.8×10^{-12} eV per molecule, far too small to have any permanent effect on the molecule or to seriously slow down the particle.

References

[1] Aharonian F A 2001 *Proc. 27th ICRC (Hamburg, August)* ed K H Kambert, G Heinzelmann and C Spiering (University of Hamburg) p 250
[2] Evans R D 1955 *The Atomic Nucleus* (New York: McGraw-Hill)
[3] Fazio G G 1967 *Annu. Rev. Astron. Astrophys.* **5** 481
[4] Fazio G G 1970 *Nature* **225** 905
[5] Fichtel C E and Trombka J I 1997 *Gamma Ray Astrophysics (NASA Ref. Publ. 1386)* p 219
[6] Frank I M and Tamm Ig 1937 *Dokl. Akad. SSSR* **14** 109

[7] Gould R J and Schreder G 1966 *Phys. Rev. Lett.* **16** 252
[8] Harwit M 1988 *Astrophysical Concepts* (Berlin: Springer)
[9] Jelley J V 1958 *Cherenkov Radiation* (New York: Pergamon)
[10] Jelley J V 1982 *Proc. Workshop on VHE Gamma Ray Astronomy (Ooty, September 1982)* ed P V Ramanamurthy and T C Weekes (Bombay: Tata Institute of Fundamental Research) p 3
[11] Jelley J V 1983 *Photochem. Photobiol. Rev.* **7** 275
[12] Lang K R 1980 *Astrophysical Formulae* (Berlin: Springer)
[13] Nikishov A J 1962 *Sov. Phys.–JETP* **14** 393

Index

active galactic nucleus (AGN), 3
afterglow, 185
AGILE, 49, 51
AMS, 50
Anasazi, 90
Andromeda Nebula, 55, 127
anti-center, 61
Apollo, 198
ASCA, 93, 95, 149–150
atmosphere, 21–22

BACODINE, 180, 185
balloon, 6, 7, 11, 42, 79, 116,
BATSE, 49, 114, 130, 170, 174, 176–177, 182–185, 188
BeppoSAX, 151, 181, 185
beryllium, 68
Big Bang, 169, 186, 192–193
black hole, 54, 112, 119, 122, 130, 156–157, 162, 185
blazar, 128–129, 132, 134,
BL Lacerate, 132, 134–135, 141–142, 194–195
bremsstrahlung, 14, 57, 62, 98, 200, 203–207

cadium zinc telluride, 48
caesium iodide, 49, 52
CANGAROO, 31, 35, 37, 85, 92–95, 128
Carter, Jimmy, 54
CASA, 38, 59
Cassiopeia A, 73, 83, 95–97
CAT, 31, 87, 97
CELESTE, 40, 53, 86

Centaurus A, 129, 140
Centaurus X-3, 112–114
CG195+4, 124
Chaco Canyon, 90
Chandra, 54, 73, 79–80
charged particle, 6
Cherenkov radiation, 7, 9, 14, 16–27, 38–40, 50, 82–83, 200, 211–214
  imaging, 28, 34
  telescopes, 58, 65, 79, 113, 123, 184
Circinus X-1, 122
Cocconi, 8, 82–83
Cold War, 170
COMPTEL, 46, 62, 84, 104, 130, 134, 139, 174, 191, 197–198
Compton Gamma Ray Observatory (CGRO), 3, 6, 43, 45, 54
Compton scattering, 5, 46–47, 57, 60, 65, 200, 203
  telescope, 46, 201
Compton-synchtrotron, 77, 84, 88–89, 93, 95–97, 130, 139, 162
Coroniti, 79
COS-B, 11, 43, 58–59, 80, 85, 96, 104, 113, 116, 124, 126, 154, 191
Cosmic Background Explorer (COBE), 195
cosmic radiation, 3, 7, 8, 13, 19, 26–28, 34, 40, 55, 57, 64–65, 69–71, 98, 113, 128,

217

133, 136, 162, 188, 206
cosmology, 2, 3, 169, 190, 193–197
Coulomb scattering, 30, 45
Crab Nebula, 9, 31, 35, 68, 73, 77–90, 92, 116, 140–141, 144, 160
Crab pulsar, 103–108, 121
Crimea, 8, 83
CTA-1, 98
Curtis, 188
curvature radiation, 109
cyclotron, 179
Cygnus A, 129, 157
Cygnus array, 38
Cygnus X-3, 9, 38, 113–114, 122, 214

Davis-Cotton, 22
diffuse, 190
Diffuse Infrared Background Experiment (DIRBE), 195–197
Doppler shift, 60, 131, 158,
Durham, University of, 72, 93, 114

EAS-TOP, 59
EGRET, xi, 43–45, 50–53, 58–59, 61, 75, 81, 84–85, 88, 92, 97, 99, 103–104, 106, 108, 111–112, 116, 120, 124, 134, 136, 141, 154, 162, 174, 191, 194, 197
electromagnetic cascade, 5, 14, 24–26
electromagnetic spectrum, 4, 13
electron, cosmic, 26, 57, 73, 77, 82–83, 88, 93–94,
Explorer XI, 11, 193
extragalactic background light (EBL), 195–199, 209
Extreme Ultraviolet Explorer, 149–150

Fermi acceleration, 71

FIRAS, 196
Fishman, Jerry, 188
Flat Spectrum Radio Source (FSRQ), 132, 134, 151, 195
fluence, 173
Fourier transform, 214
Frank, 213

Galaxy, 55–56, 60–61, 64, 69–71, 126, 180, 193
galactic center, 56, 61, 65, 180
galactic halo, 68
galactic plane, 55–56, 60–61, 65, 116, 180
Gamma Cygni, 98
Geminga, 104–108, 119, 121, 124
germanium, 48
GLAST, 49–54, 75, 111, 120, 140, 194
Goddard, 174, 180
Gould, 84, 214
Gould's Belt, 118–119, 121
GRAAL, 86
GRB910503, 182
GRB940217, 183–184
GRB970228, 181
GRB970508, 181
GRB980425, 182
GRB990123, 180, 185
GRS1915+105, 122
GT-48, 31
GT0236+610, 120
G312.4-0.4, 98
G40.5-0.5, 98

hadrons, 71, 89, 95
Havarah Park, 113–114
HBL, 132, 164
HEGRA, 31, 35, 58–59, 87–88, 95–97, 123, 144, 146,
Hercules X-1, 114
HESS, 35–36, 75
High Energy (HE), 4
Hill, 39

Hillas, A M, 14
Hubble telescope, 54, 76, 78–79, 209
Huygens's wavelet, 212
hydrogen, 60
hypernova, 186, 188
H1426+428, 142–144, 199

IC433, 98–99
image intensifier, 39
IMP-6, 174
INTEGRAL, 48
interstellar dust, 55
interstellar gas, 55–56, 60, 62, 95

Jelley, John, xii, 211, 213

Kennel, 79
Kepler, 68
Kes67, 98
Kiel, 113–114
Klein–Nishina, 83, 88, 161, 202

Lamb, Don, 188
Large Magellanic Cloud (LMC), 67, 72, 75, 127–128, 133
Larmor precession, 209
LBL, 132, 164
Lebedev Institute, 83
LMC X-1, 114
Local Group, 127, 191
Lord Rosse, 77
Lorentz factor, 158–160, 186, 205
Los Alamos, 172
LSI+61-303, 120

MAGIC, 34–37, 53
magnetic stars, 69
Markarian 421 (Mrk421), 136, 140–151, 165, 197, 199
Markarian 501 (Mrk501), 140–143, 149, 152, 158, 165, 197, 199
Medium Energy (ME), 4
meteor, 16, 22
microquasar, 113, 122

microwave background, 2, 214
Milagrito, 38, 124, 185
Milagro, 38, 53, 86, 185
Milky Way, 57, 126
molecular clouds, 57, 60–61, 72, 98, 118
moment of inertia, 102
Monoceros, 98
Monte Carlo, 14, 16–17, 26
Morrison, Philip, 5
M31, 179
M82, 128
M87, 168

NaI(Tl), 45
Narrabri, 93
neutralino, 122
neutrinos, 68, 72, 75
neutron star, 95, 109, 112, 178, 180
NGC253, 128
night-sky, 19
Nikishov, 214
nova, 69

OB stars, 64, 116, 124
OJ+287, 136
Oort Belt, 177
Orion, 60
OSO-3, 64–65
OSSE, 104, 128, 130, 174
outer gap, 109

Paczynski, Bohdan, 188
pair production, 5, 14, 42–43, 143, 146, 158, 184, 195, 200, 205–209
particle acceleration, 3
photomultiplier (PMT), 19, 22, 26, 40, 46, 52
pictograph, 90
pion, 62, 74, 82, 94, 97–98, 163, 200, 203–206
plerion, 73, 92, 160
PKS0528+134, 135–136, 154, 161, 164

PKS1331+170, 136
PKS1622-297, 135
PKS2155-304, 136
polar cap, 109
polarization, 3, 26, 134, 158, 209
Porter, Neil, xii, 39–40
primordial black holes (PBH), 178, 193
PSR B1055-52, 104–108
PSR B1509-58, 104, 111
PSR B1706-44, 92–93, 104
PSR B1951+32, 104–108, 111
pulsar, 75, 77, 80–81, 92,
   accretion-powered, 102
   millisecond, 102
   rotation-powered, 102

Quantum efficiency, 22–24
quasar, 130–132, 140

radiation length, 13–14
radio astronomy, 13
radio galaxies, 8
RBL, 132
redshift, 137, 143
Rees, Martin, 166
refractive index, 211
relativistic, 2
   jets, 3, 130, 157–160
Rho Oph, 61
ROSAT, 73, 79, 92
ROTSE, 180
RXJ1713.7-3946, 95–96
RX1836.2+5925, 121

SAS-2, 9, 11, 43, 58, 80, 97, 113, 116, 124, 191, 198
SAXJ0635+0533, 121
scintillator, 45–46, 52, 65, 171
Sco X-1, 11, 114
Sedov phase, 74–75
Seyfert, Carl, 129
SHALON, 31
Shapley, 188

shock waves, 70
silicon strip, 48–49, 51
SIRTF, 54
Small Magellanic Cloud (SMC), 57, 127–128, 133
Smithsonian Astrophysical Observatory, 83
Smithsonian Museum, 188
SN1006, 68, 73, 93–94, 96, 99
SN1572, 68
SN1604, 68
SN1987a, 67–68, 72, 75
SN1991T, 154
SN1998bw, 182
Solar System, 1, 60–61, 65, 70, 178
Solar Maximum Mission (SMM), 169, 198
Solar Two, 38, 86
Southampton, University of, 79
Space Station, 50
spark chamber, 11, 42, 45, 79, 116
spectral energy distribution (SED), 139
spirals, 126
SS433, 122
STACEE, 39, 53, 86
Steady State, 192
Sullivan, Walter, 54
supercluster, 127
superluminal, 134, 158, 166
supernova, 3, 8, 57, 70, 174
   type Ia, 67–68
   type II, 67–68, 75
supernova remnants (SNR), 64
supranova, 186
Swift, 49
synchrotron, 3, 82, 200, 209–210
S147, 98

TACTIC, 30
Tamm, 211
thermal processes, 2
Thompson scattering, 161, 203–205
TIBET, 59, 86–87

transition radiation, 50
Tycho, 68, 73, 75, 99

Ultra High Energy (UHE), 4

Vavilov, 213
Vela, 93, 116
   pulsar, 104–108
   satellite, 171, 173, 175
   SNR, 73
   X-1, 114
VERITAS, 37, 53, 75
Very High Energy (VHE), 4
Virgo Cluster, 127, 191
VLA, 122
VLBI, 122, 166

W Comae, 136
Whipple, 9, 30–31, 53, 58–59, 75, 81, 83, 86–87, 97, 99, 123, 142, 149, 150–151
white dwarf, 67, 112
Wolf–Rayet star, 96
W28, 98
W44, 98–99
W51, 99

W63, 99

XBL, 132
x-ray astronomy, 2, 11

zodiacal light, 191

1ES1740.7-2942, 122
1ES1959+650, 142–144, 199
1ES2344+514, 142–144
2CG135+01, 120
3C66A, 136, 142
3C175, 129
3C273, 116, 126, 136, 141–142, 154, 158
3C279, 136–137, 142, 154, 159
3EG J1714-3857, 95
3EG J0241+6103, 120
3EG J0634+0521, 121
3EG J1324-4314, 130
3EG J1744-3039, 121
3EG J1824-1514, 122
3EG J2033+4118, 123
4U0115+63, 114
4U0241+61, 120